AF800587

Risk Vigilant Bharat

This handbook is dedicated to the countless process engineers and operational experts I have had the privilege of engaging with over the past two decades. Your patience, generosity, and willingness to share technical insights have been the bedrock of my education in this industry.

Preface

Mining is a business of paradoxes. It is an industry where crores in capital investment sit on the books, yet the daily profitability is determined by the specific gravity of ore, the efficiency of a single blast, or the integrity of a weighbridge operator at 3:00 AM.

For an Internal Auditor, walking into a mine site can be an intimidating experience. The sheer scale of operations—the roar of Heavy Earth Moving Machinery, the complex chemistry of beneficiation plants, and the distinct language of geologists—often creates an invisible wall between the "Office" and the "Pit."

For too long, this disconnect has forced auditors to retreat into their comfort zones. We stick to the safety of the administrative block, verifying vouchers, checking purchase orders, and ticking compliance boxes. The result? "Generic Observations." We report on filing errors and authorization delays, while the real value—the massive leakages in stripping ratios, yield losses, and logistics theft—remains buried under the dust of operations.

I wrote this handbook to dismantle that wall.

Over the last two decades practicing Risk-Based Internal Audit (RBIA), I have learned that you cannot audit a mine from a spreadsheet. You have to audit it from the ground up. However, I also realize that not every auditor has the luxury of spending years in the field to learn the nuances of geology, metallurgy, and fleet management.

This book is the accelerator for that learning curve.

It is designed as a tactical weapon for the field auditor. It distils 2 decades of "dirt-under-the-fingernails" experience into actionable audit programs. It translates complex engineering concepts—like Station Heat Rates, Stripping Ratios, and Ore Recovery—into the language of financial risk and control.

Whether you are a young auditor stepping onto a mine site for the first time, or a Chief Audit Executive looking to pivot your function from "Compliance" to "Value Addition," this book is your roadmap.

The objective is simple: To equip you with the insights required to stop being a spectator in the administrative block and start being a value-creator in the field.

Welcome to the pit.

Aditya Pratap Singh
aditya.pratap.singh@icai.org
January 2026

My Linkedin Profile

Master Index

Part I- The Foundation (Industry & Strategy)

1.1 Introduction: The Business of Dirt

For an auditor, the metal mining industry in India is unique. Unlike manufacturing, where you buy raw materials and add value, in mining, you find the raw material (Exploration), fight for the right to dig it up (Regulatory Auctions), and then move it. The value doesn't come from "creating" a product; it comes from extracting it efficiently and proving its quality.

The Post-2015 Shift: Since the amendment of the MMDR Act in 2015 (and further in 2021), India has moved from "Allocation" to "Auction". Companies now pay huge premiums (sometimes >100% of the sale value) to the government just for the right to mine. This has squeezed margins and made cost control and volume maximization the primary audit objectives. This also created a massive incentive for cost suppression or illegal mining to break even.

1.2. The Mining Value Chain (The Audit Map)

To audit a mine, the auditor must visualize the physical flow.

a) **Discovery**

Activity: Finding the ore body and estimating "Reserves" (Proven/Probable).

Key Process: Identifying the resource location and quality before any digging begins.

b) **Development**

Activity: Building roads, stripping the topsoil (Overburden), and setting up the pit.

Key Process: Mine Planning and Pre-Stripping to expose the ore body.

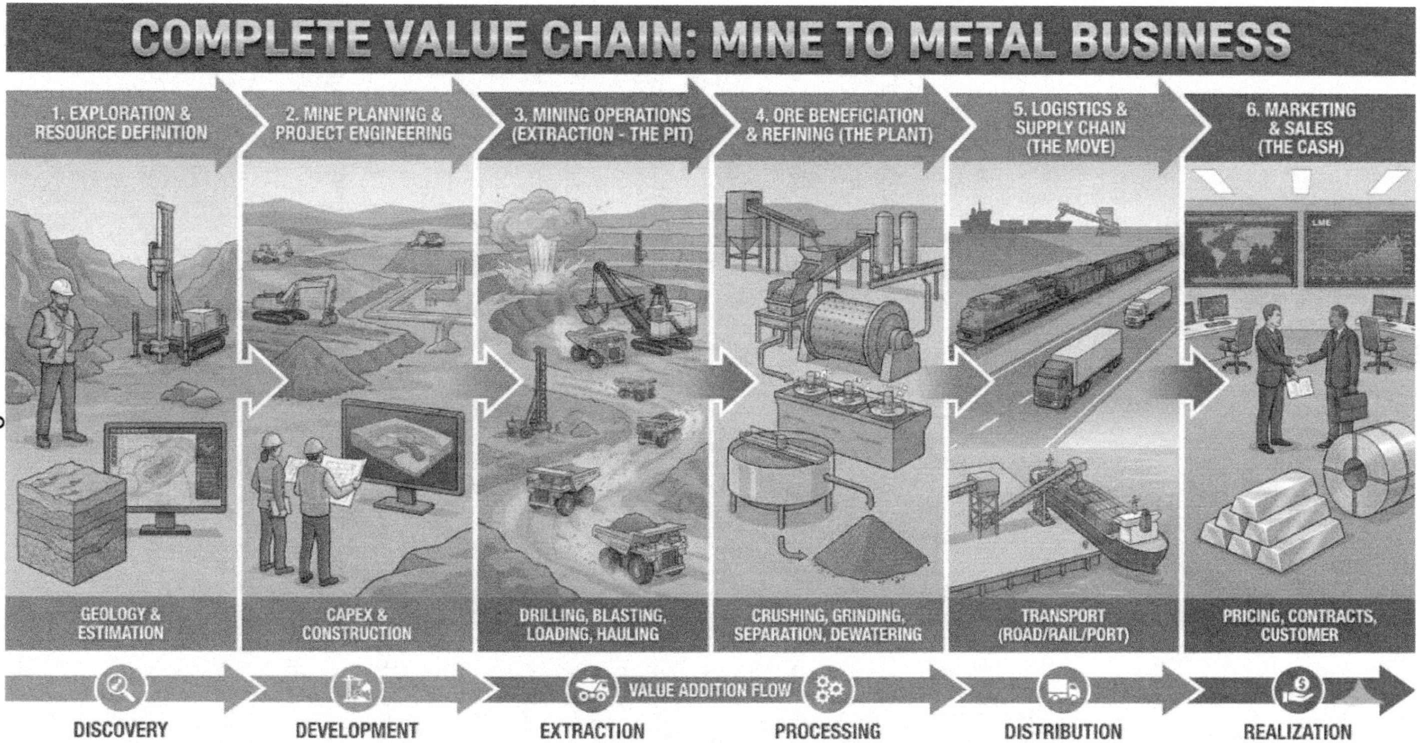
COMPLETE VALUE CHAIN: MINE TO METAL BUSINESS
1. EXPLORATION & RESOURCE DEFINITION
2. MINE PLANNING & PROJECT ENGINEERING
3. MINING OPERATIONS (EXTRACTION - THE PIT)
4. ORE BENEFICIATION & REFINING (THE PLANT)
5. LOGISTICS & SUPPLY CHAIN (THE MOVE)
6. MARKETING & SALES (THE CASH)
LME
GEOLOGY & ESTIMATION
CAPEX & CONSTRUCTION
DRILLING, BLASTING, LOADING, HAULING
CRUSHING, GRINDING, SEPARATION, DEWATERING
TRANSPORT (ROAD/RAIL/PORT)
PRICING, CONTRACTS, CUSTOMER
VALUE ADDITION FLOW
DISCOVERY
DEVELOPMENT
EXTRACTION
PROCESSING
DISTRIBUTION
REALIZATION

c) **Extraction**

Activity: The physical removal of ore from the ground.

Process Flow: Drill Planning > Blasting > Excavation (Loading) > Hauling > Crushing > Stockpiling.

d) **Processing / Beneficiation**

Activity: Crushing, Screening, and Washing the raw ore to improve its grade.

Key Process: Converting Run of Mine (ROM) ore into Concentrate by separating the valuable metal from the waste (Tailings).

e) **Distribution**

Activity: Evacuation of the finished product to the customer/intermediary plant.

Process Flow: Transporting via Road, Rail (Rake), or Slurry Pipeline to steel plants or ports.

f) **Realisation**

Activity: Selling the final product to customers (e.g., Steel Plants/Smelters) or captive plant.

Key Process: Pricing based on global indices and managing commercial contracts.

1.3. The Profit Equation: What Moves the Needle?

Understanding this equation helps the auditor identify Significant Accounts and Risks.

Profit = (Volume x Grade x Market Price)- (Royalty + Logistics + Extraction Cost)

Mapping of these "Profit Variables" with some typical risks and fraud schemes prevalent in industry.

Profit Variable	Risk Sub-Category	Detailed Risk / Fraud Scheme
1. VOLUME: The quantity of ore excavated and sold.	Weighbridge Manipulation	The Water Tanker Trick: Trucks enter with hidden water tanks to increase tare weight, then dump water before loading ore to steal the equivalent weight in mineral.
		Sensor Tampering: Use of remote-controlled devices ("chips") to alter digital weight readings by 5-10%.
		Axle Skipping: Positioning the truck so one axle is off the weighbridge plate during weighing to under-report dispatch weight.
	Stockpile Manipulation	Density Inflation: Using an inflated "Bulk Density" factor to convert volumetric survey (m3) into tonnage, creating "paper tonnes" that don't exist physically.
		The Carpet Trick: Spreading the stockpile over a large area (low height) to make volumetric measurement difficult and hide shortages.
	Dispatch Fraud	Bypass Dispatch: Trucks leaving the mine via unauthorized exit routes without crossing the weighbridge.
2. GRADE: The quality (e.g., Fe % in Iron Ore).	Classification Fraud	Screen Manipulation: Intentionally using larger screen meshes to allow high-value "Lumps" to fall into the "Fines" pile, which attracts lower royalty and sale price.
	Sampling & Lab Fraud	Salting: Adding high-grade material to exploration or dispatch samples to artificially boost the reported grade.
		Selective Sampling: Taking samples only from the "best" part of the heap rather than a representative cone-and-quarter mix.

		Moisture Arbitrage: Manipulating moisture content readings. Overstating moisture reduces the payable "Dry Metric Tonne" weight, causing revenue loss.
3. MARKET PRICE: Driven by global indices (Platts, ODEX) and IBM ASP.	Pricing Application	Index Manipulation: Applying the wrong price index date (e.g., using last month's lower index for a related party sale) instead of the contractual dispatch date index.
		Discount Schemes: Offering unauthorized "quality discounts" or "wet cargo allowances" that are not backed by third-party lab reports.
	Related Party Transactions	Transfer Pricing: Systematically under-pricing ore sold to a sister company (steel plant) to shift profits to a lower-tax unit.
4. ROYALTY: Paid to State Govt, DMF, NMET.	Calculation Errors	Base Price Error: Calculating royalty on the *Invoiced Price* rather than the *IBM Published Average Sale Price* (ASP), especially when IBM ASP is higher (leading to statutory short-payment).
		Sizing Misdeclaration: Reporting production as 80% Fines (lower royalty) and 20% Lumps (higher royalty) when actual crusher output is 50:50.
5. LOGISTICS: Cost of moving ore.	Freight Leaks	Dead Freight: Paying for full wagon capacity (e.g., 65 tonnes) but loading only 60 tonnes due to poor belt calibration.
		Punitive Overloading: Systematically overloading trucks/wagons, attracting 6x penalty rates from Railways or Transport Departments.
	Theft in Transit	Grade Swapping: High-grade ore dispatched from the mine is swapped for low-grade dust at an unauthorized roadside yard during transit.

		GPS Spoofing: Trucks creating "fake trips" or deviating from geo-fenced routes to steal fuel or cargo.
6. EXTRACTION COST: Fuel, Explosives, Spares	Fuel Theft	Siphoning: Direct theft of diesel from excavators and dumpers.
		Idling Fraud: Keeping machines running (idling) to burn fuel and increase "Hour Meter Readings" without doing actual work.
	Explosives Pilferage	Fake Blasting: Diverting explosives to illegal markets while claiming they were used for "secondary blasting" or hard rock strata.
	Stripping Ratio (SR)	Capitalization Fraud: Classifying routine operational waste removal as "Capital Pre-Stripping" to inflate current profits.

1.4. Key Stakeholders (The Regulator's Watch)

An auditor in India is not just checking internal policy; they are checking compliance with these external bodies. A failure here shuts down the mine.

1. IBM (Indian Bureau of Mines): The technical regulator. They approve the *Mining Plan* (how much you can dig) and monitor the *Grade*.
2. State Department of Mines & Geology: They collect the Royalty. They issue the *Transit Passes* (e-Pass) required for every truck.
3. MoEF&CC (Ministry of Environment): They issue the *Environmental Clearance (EC)*. Breaching the EC production limit is a criminal offense.
4. DGMS (Directorate General of Mines Safety): They govern safety. Accidents can lead to immediate suspension of operations.

1.5. The "Why" Box: Auditor's Mindset

Auditor's Insight: The "Invisible" Inventory

In a retail audit, you can count the shirts on the shelf. In mining, your inventory is underground or in massive piles that change shape daily.

- The Challenge: You cannot "count" a 50,000-tonne stockpile with a clipboard. You need surveyors and drones.
- The Mindset: Don't trust the spreadsheet. Trust the *physics*.
    - If they claim they mined 100,000 tonnes, did they burn enough diesel to move that much earth?
    - If they claim the ore is Low Grade, does the geological map of that pit area support that claim?
    - Auditors must vigilantly check for 'off-book' mining (mining outside the lease area) or grade suppression to artificially lower the premium payable base.

Your job is to correlate Operations Data (GPS, Fuel, Blasting) with Financial Data.

1.6. Captive vs. Merchant: A Captive mine's objective is 'Security of Supply' (Reliability), while a Merchant mine's objective is 'Profit Maximization' (Price).

Audit Impact: For Captive mines, the 'Transfer Price' to the plant is a major audit risk for tax evasion (under-pricing to shift profit to the plant)."

1.7. Commodity Nuances: One Size Does Not Fit All

An auditor cannot audit a Coal mine with an Iron Ore checklist. While the machinery is similar, the Fraud Risks and Value Drivers are completely different. While the act of digging is universal, the audit logic changes based on what is coming out of the ground. Below are the specific characteristics and risks for India's major mined commodities.

A. Iron Ore (The Steel Feeder)

- Context: The backbone of the steel industry. Found mostly in Odisha, Karnataka, Chhattisgarh, and Jharkhand.
- Value Driver: Fe Content (Iron %) and Physical Form (Lumps vs. Fines).
    - *Lumps (Rocks):* High value, used directly in Blast Furnaces.
    - *Fines (Dust):* Lower value, needs sintering/pelletizing.
- Key Audit Difference:
    - The "Screening" Fraud: The biggest risk is classifying Lumps as Fines. Lumps attract higher royalty and market price. Operators may manipulate the screen size to let small lumps pass through as fines to save money.
    - Moisture: Iron ore is porous. During monsoon, "Blue Dust" absorbs massive water weight. Auditing the "Dry Weight" conversion is critical.

B. Coal (The Energy Source)

- Context: Used for Power (Thermal Coal) and Steel (Coking Coal). "Black Gold" is the most pilfered commodity in India.
- Value Driver: GCV (Gross Calorific Value). Coal is sold by "Energy Content" (Kcal/kg), not just weight.
- Key Audit Difference:
    - Grade Slippage: The supplier bills for G8 (High quality) but delivers G12 (Low quality/High Ash). The auditor must check the Third-Party Sampling (CIMFR) results at the loading point vs. the unloading point.
    - Reject Mixing: Mixing shale/stones (black rocks that look like coal but don't burn) to increase weight.
    - Spontaneous Combustion: Coal piles catch fire if left too long. Audit the "First-In-First-Out" (FIFO) logic in stockpile management.

C. Limestone (The Cement Binder)

- Context: Low-value, high-volume mineral mostly captive to Cement plants.
- Valuation Driver: CaO (Calcium Oxide) % and Silica content.
- Key Audit Difference:
    - Blending Logic: Since it's low value, theft isn't the main risk. The risk is efficiency. The audit focuses on the "Blending Ratio" at the crusher—mixing high-grade limestone with low-grade to optimize the mine life without ruining the cement quality.
    - Explosive Ratio: Limestone is softer than Iron Ore. If the Powder Factor (explosives used per ton) is high, it indicates inefficiency or theft of explosives.

D. Base Metals (Zinc, Lead, Copper)

- Context: Often deep underground mines (e.g., Hindustan Zinc). High value, low volume.
- Value Driver: Metal Content % (e.g., 6% Zinc in ore).
- Key Audit Difference:
    - Underground Safety: The audit includes high-cost items like Ventilation and Strata Control (roof bolting).
    - Mill Recovery: Since the rock is 94% waste and 6% metal, the "Recovery Rate" in the mill is everything. A 1% drop in recovery is a massive financial loss.

The "Commodity Matrix" for Auditors

Feature	Iron Ore	Bauxite	Coal	Limestone
Primary Unit	Wet Metric Tonne (WMT)	Tonne	Tonne (GCV Adjusted)	Tonne (CaO Adjusted)
Royalty Basis	Ad Valorem (IBM ASP)	Linked to LME	Fixed + % (Price of Coal)	Fixed Rate per Tonne

		Aluminium Price		
Key Quality Param	Fe %	Alumina (Al2O3) % and Reactive Silica	GCV (Kcal/kg)	CaO %
Highest Fraud Risk	Grade Suppression (Lump \to Fine)	Land Acquisition (Tribal areas) & Logistics	Grade Slippage & Stone Mixing	Waste Dumping as Production
Logistics Focus	Rake Loading Limit (Overloading penalty)	Detention / demurrage charges. And over billing by Contractor.	Shortage/Theft in Transit	Internal Conveyor Belts

1.6. Common Industry Jargon (Glossary)

- ROM (Run of Mine): Raw, unprocessed ore directly from the blasting face.
- Overburden (OB): The waste rock/soil overlying the ore body.
- Stripping Ratio: The ratio of Waste to Ore (e.g., 3:1 means you dig 3 tons of waste to get 1 ton of ore). *A higher actual stripping ratio than planned destroys profitability.*
- WMT vs. DMT: Wet Metric Tonne (with moisture) vs. Dry Metric Tonne (without moisture). *Pricing discrepancies often hide here.*
- Rake: A full goods train (usually 58 wagons).

For an auditor, this is the most critical chapter for Compliance Testing **(MMDR, IBM, DGMS, & Environment)**. In India, a violation here doesn't just mean a fine; it may mean cancellation of the Mining Lease (ML) and criminal proceedings against the Directors (and the Occupier).

2.1. Introduction: The License to Operate

In the Indian context, a mining company does not "own" the land in the traditional sense; they are a "Lessee" of the State Government. The regulatory web is complex because it involves three distinct layers of authority:

1. The Owner: State Government (Grants the Lease, Collects Royalty).
2. The Technical Regulator: Indian Bureau of Mines (IBM) for Major Minerals; Coal Controller (CCO) for Coal.
3. The Safety/Enviro Regulator: DGMS (Safety) and MoEF&CC (Environment).

The Audit Objective: To ensure the "License to Operate" remains valid and that statutory payments are accurate.

2.2. The Constitution: MMDR Act, 1957 (and Amendments)

The *Mines and Minerals (Development and Regulation) Act* is the parent act. As an auditor, you must focus on Section 21(5)—the penalty clause. Illegal mining (mining without license or outside the lease area) attracts a penalty of 100% of the value of the mineral, plus jail time.

Key Audit Checks (The "Must-Haves"):

- The Mining Lease (ML) Deed: Verify the validity period. (Post-2015 auction regime, leases are usually for 50 years). "Deemed Extension" clause for older leases (Section 8A of MMDR Act), is a common area of litigation. For pre-2015 leases, verify the 'Supplementary Lease Deed' extending the period to 2030 (Merchant) or 2035 (Captive). Ensure the State Govt has formally executed this.

- Surface Rights: The "Mining Lease" gives you the right to the mineral underground. It does *not* automatically give you the right to the land surface. Check if the company has acquired the Surface Rights from private land owners or the Govt to enter the land.

2.3. The Technical Rulebook: MCDR, 2017

The *Mineral Conservation and Development Rules (MCDR)* govern *how* you mine. The Bible here is the "Approved Mining Plan."

Key checkpoints (MCDR, MMDR Act, Mines Act)

Statute	Compliance Requirement	Audit Procedure	Red Flag / Risk
MMDR Act: Sec 4(1)	Adherence to Mining Plan	Compare Annual Production Quantity in SAP vs. limit in Approved Mining Plan.	Over-production: Extracting >100% of plan = 100% penalty on excess value.
MMDR Act: Sec 21(5)	Illegal Mining (Boundaries)	Overlay GPS coordinates of pit boundaries onto Lease Map.	Encroachment: Excavation extending outside lease (even by meters) is illegal.
MMDR Act: Sec 9(2)	Royalty Payment	Re-calculate Royalty: Dispatch Qty * IBM Avg Sale Price * Rate %. Match with Challan.	Short Payment: Paying on "Pit Head Value" instead of "IBM ASP".

MMDR Act: Sec 9B/9C	DMF & NMET	Verify DMF (10% or 30%) and NMET (2%) are calculated on the *Royalty* amount.	Rate Error: Applying 10% rate (Auction) to a legacy mine (30% liability).
MMDR Act: Sec 8A	Lease Validity	Check Supplementary Lease Deed for extension endorsement (to 2030/2035).	Expired Lease: Operating on "Deemed Extension" without formal deed.
MCDR 2017: Rule 11(1)	Scientific Mining	Compare Actual Stripping Ratio (Waste/Ore) vs. Planned Ratio.	High-Grading: Mining ore but ignoring waste removal to boost short-term profit.
MCDR 2017: Rule 45	Statutory Returns	Reconcile Closing Stock in Form F-1/G-1 with Audited Balance Sheet.	Stock Suppression: Under-reporting stock to Govt to hide theft/evade royalty.
MCDR 2017: Rule 12(4)	Exploration Scheme	Check Drilling Logbook for committed boreholes (G3 to G1 upgrade).	Lease Termination: Failure to explore risks cancellation of lease.
MCDR 2017: Rule 35	Star Rating	Validate Self-Certified Star Rating against physical site (ETP, Walls).	False Declaration: Claiming "5-Star" to get clearances while violating norms.

MCDR 2017: Rule 33	Topsoil Preservation	Inspect Topsoil Stacks. Must be stored separately with signage/vegetation.	Resource Destruction: Dumping waste rock on fertile topsoil.
Mines Act: Sec 17	Manager Appointment	Verify First Class Manager's Certificate and physical presence.	Ghost Manager: Operating with foreman using remote manager's certificate.
Mines Act: Sec 18	Owner's Duty (Finance)	Review "Safety Budget". Ensure no safety requisition was rejected by Finance.	Deemed Owner Liability: Directors are criminally liable if they refuse safety funds.
Mines Act: Sec 30/31	Working Hours	Analyze Biometric data for "Double Shifts" (16 hrs) & Overtime violations.	Fatigue Risk: Systematic 48-hour week violation leading to accidents.
Mines Act: Sec 48	Employee Register	Physical head-count in pit vs. Form A/B Register names.	Ghost Workers: Names in register not on site (or vice versa).

2.4. The Financials: Royalty, DMF, & NMET

This is where the money is.

- Royalty: Paid to State Govt. (e.g., 15% of Sales Price for Iron Ore).
- DMF (District Mineral Foundation): Used for local development. 30% for leases granted before 2015 (Old); 10% for leases granted via Auction (New).

- NMET (National Mineral Exploration Trust): 2% of Royalty.

The Audit Formula: Total Liability = (Qty x ASP x Royalty Rate) + (Royalty x 0.30) + (Royalty x 0.02)

(Where ASP = Average Sale Price published by IBM)

Fraud Alert: The "Size" Game. Royalty rates often differ by size.

- *Example:* Lumps = High Royalty. Fines = Low Royalty.
- *The Audit Step:* Check the Crusher Production Log. If the crusher is set to produce 60% Lumps, but the Royalty Return shows 80% Fines, the company is mis declaring production to save tax.

2.5. Safety & Environment (DGMS & MoEF)

A. DGMS (Directorate General of Mines Safety)

Governed by *The Mines Act, 1952*.

- The Appointment Rule: A mine with > X production *must* have a First Class Mines Manager (FCC).
    - *Audit Check:* Verify the FCC certificate of the Mines Manager. If the FCC holder is on leave, who handled the charge? (Operating without a qualified manager is a statutory breach).
- Working Hours: No worker should work >48 hours/week (Section 30).
    - *Audit Check:* Pull the "Time Office" records. Look for "Double Shifts" (OT) being disguised.

B. Environment (The EC Limit)

- EC (Environmental Clearance): Sets the maximum Cap (e.g., 4 Million Tonnes Per Annum- MTPA).
- CTO (Consent to Operate): Issued by the State Pollution Control Board (PCB). Needs annual renewal.
    - *Audit Check:* Ensure the CTO is valid for the audit period. Check water consumption logs and other aspects against the limit in the CTO.

2.6. Sector Specifics: The Regulatory Nuances

Feature	Coal Mining	Non-Coal (Iron/Limestone/Bauxite)
Primary Regulator	Coal Controller Organization (CCO)	Indian Bureau of Mines (IBM)
Governing Act	Coal Mines (Special Provisions) Act, 2015	MMDR Act, 1957
Royalty Basis	Mixed (Fixed amount + % of price)	Mostly Ad Valorem (Based on IBM ASP)
Dispatch Permit	Mineral Dispatch Permit (MDP)	e-Transit Pass (Integrated with State Portal)
Safety Zone	Higher focus on Gas/Fire (Inflammable)	Higher focus on Slope Stability (Landslide)

2.7. The "Why" Box

Auditor's Insight: Why check the "Form G-1"? The *Form G-1* is the Annual Return filed with the government.

- The Trap: Companies often under-report Closing Stock in Form G-1 to show lower inventory handling, But they report higher Closing Stock to banks to get higher Working Capital limits.
- The Check: Compare Closing Stock (Form G-1) vs. Closing Stock (Bank Stock Statement) vs. Closing Stock (ERP).
- The Result: If they don't match, you have found either *Statutory Fraud* or *Bank Fraud*.

Chapter 3: Geology & Exploration (The Search)

For many auditors, this area feels "too technical" to touch. As a result, it becomes a black box where millions of rupees are "buried" without scrutiny. This chapter demystifies the process, focusing on the commercial and control aspects of exploration.

3.1. Introduction: The High-Stakes Gamble

Exploration is the R&D of mining. It involves spending cash today to find assets (Ore) that might be mined 10 years later.

- The Accounting Standard: Ind AS 106 (Exploration for and Evaluation of Mineral Resources).
- The Goal: To move confidence from "G4" (Reconnaissance/Guess) to "G1" (Proven Reserve).
- The Cost Driver: Drilling. It costs thousands of rupees per meter to drill into the earth.

The Audit Objective: To ensure that drilling costs are genuine (work actually done), accurately classified (Capex vs. Opex), and that the geological data is secure.

3.2. Process Notes:

3.2.1 The Exploration Lifecycle

- Survey: Satellite/Aerial mapping to identify potential zones.
- Drilling: Bringing in Rigs (Diamond Core or Reverse Circulation) to bore into the ground.
- Core Logging: Geologists examine the cylindrical rock samples ("Cores") extracted from the ground and record rock type, structure, and visible mineralization.
- Assaying: Sending core samples to the lab for chemical analysis.
- Modeling: Using software (Surpac/Datamine) to build a 3D model of the ore body.

- Geological Report Output: The exploration data determines the Life-of-Mine Stripping Ratio (Waste:Ore). The Auditor must 'freeze' this ratio here. *Any deviation in Chapter 5 (Mine Planning) must be cross-referenced back to this Geological Report.*

3.2.2. The Core Box: The Physical Evidence

While the "Geological Report" is just paper, the Core Box is the reality. This is the auditor's primary physical proof that exploration actually happened.

When a Diamond Drill (DD) machine bores into the earth, it extracts a solid cylinder of rock called a Core. This core is the only physical proof of the ore body's existence. These cores are stored in partitioned trays called "Core Boxes," which serve as the permanent library of the mine's assets.

Key Components of a Core Box:

1. The Core: The cylindrical rock sample itself.
2. The Dividers: Separate the core runs (usually 1-meter lengths).
3. The Run Markers: Small wooden or plastic blocks placed by the driller to mark the depth (e.g., "Depth: 53.0m"). These are crucial for calculating "Core Recovery."
4. The ID Plate: The outside of the box must list the Borehole ID, Box Number, and Depth Range (e.g., *BH-2024-05 / Box 4 of 10 / 12m to 16m*).

Audit Check: The "Core Library" Walkthrough

Audit Checkpoint	Red Flag / Risk	Procedure
Physical Existence	Missing Boxes: "We lost them in the rain" or "eaten by termites" is a major red flag. If the core is gone, the "Proven Reserve" status is invalid.	Select 3 Boreholes from the "Geological Report" (e.g., BH-105). Ask the Geologist to retrieve Box #1 and Box #Last for that hole.
Core Recovery %	Low Recovery: If the box has only 2 meters of rock for a 3-meter drilled depth (66% recovery), the assay grade will be misleading.	Measure the length of rock inside a 3-meter section of the box.

Sampling Integrity	Empty Box: If the *entire* core was sent to the lab, there is no "Counter Sample" left for verification.	Look for the "Split Line." Half the core should remain in the box for reference; the other half is sent to the lab.

3.2.3: Understanding the "G-Axis": The Confidence Ladder (G4 to G1)

For an auditor, "Mineral Reserves" are simply Inventory that hasn't been extracted yet. However, unlike a warehouse where you can physically count boxes, underground inventory is an estimate based on probability. In India, the Indian Bureau of Mines (IBM) mandates the UNFC (United Nations Framework Classification) system to grade this probability.

Think of this as the "Resolution" of a photograph.

- G4 is a blurry shape (It might be a rock, it might be a treasure).
- G1 is a 4K High-Definition image (We know exactly where the gold is).

As an auditor, your job is to ensure the company isn't valuing "Blurry Shapes" (G3/G4) as "Bankable Assets" (G1).

The Four Levels of Geological Confidence:

G4: Reconnaissance (The Guess)

- What it is: The geologist has walked the land, seen some surface outcrops, and maybe taken a few rock chips.
- Audit Reality: Pure speculation. No drilling has been done.
- Financial Impact: Value = Zero. Cannot be booked as an asset.

G3: Prospecting / Inferred (The Hint)

- What it is: Wide-spaced drilling (e.g., boreholes 400 meters apart). We know mineralization exists, but the continuity is assumed, not verified.
- Audit Reality: High Risk. The ore might "pinch out" (disappear) between two drill holes.
- Financial Impact: cannot be used for Mine Planning. Banks usually do not lend money against G3 resources.

G2: General Exploration / Indicated (The Estimate)

- What it is: Closer drilling (e.g., 100-200 meters apart). The shape and grade of the ore body are reasonably known.

- Audit Reality: "Probable Reserves." You can build a Pre-Feasibility Study (PFS) on this.
- Financial Impact: Acceptable for valuation, but with a risk discount.

G1: Detailed Exploration / Proved (The Fact)

- What it is: Dense drilling (e.g., 50 meters apart or less). We know the length, width, depth, and grade with 90%+ confidence.
- Audit Reality: "Proved Reserves." This is the only category that allows for a "Bankable Feasibility Study" (BFS).
- Financial Impact: This is the hard asset on the books.

The Auditor's Confidence Matrix

Code	Stage	Activity Required	Confidence Level	Audit Implication
G1	Detailed Exploration	Close-grid drilling + Pit sampling	High (>90%)	Bankable. Can be used for Amortization calculations and Reserve Audits.
G2	General Exploration	Medium-grid drilling	Moderate (~70%)	Probable. Used for long-term Life of Mine (LOM) planning but needs upgrading.
G3	Prospecting	Wide-spaced drilling	Low (<50%)	Speculative. Do NOT allow the client to use G3 tons in the "Recoverable Value" calculation.
G4	Reconnaissance	Surface mapping / Aero-survey	Very Low	Intangible. Just an expense; no asset created yet.

3.2.3: The Audit Trap: The "Inferred" Inflation

The Fraud: Management often includes G3 (Inferred) resources in their "Total Reserves" statement to investors or banks to make the mine look richer and last longer.

The Test:

1. Request the "Geological Report" (GR).
2. Look for the table breakdown of Reserves.
3. Subtract G3 from the total.
4. *Audit Query:* "Why is the Amortization rate based on (G1+G2+G3)? It should strictly be based on (G1+G2). Please recalculate the depreciation."

3.3. The "Why" Box

Auditor's Insight: The "Phantom Meter" Fraud

3.3.1 Drilling contractors are paid per meter of depth.

- The Risk: A contractor might drill 80 meters but bill for 100 meters. Or they might drill through "Soft Soil" (cheap/fast) but bill for "Hard Rock" (expensive/slow).
- The Check: You cannot measure the hole after it's drilled (it often collapses). You must audit the Physical Core. If they billed for 100m, there must be 100m of rock in the Core Box.
- The Red Flag: "100% Core Recovery" in fractured zones. (It is geologically impossible to recover every inch of rock in loose ground).

3.3.2 Land Access, Community Consent & Social License

Before the first rig is mobilised, the company needs not only geological prospects but also a "social license" to enter and work on the land.

- Land Tenure & Access Rights
    - Audit Check: Verify that surface access has been lawfully obtained from private landowners, panchayats or government agencies before drilling is started (MoUs, lease deeds, easement agreements).

    - Red Flag: Drilling carried out on "verbal permission" with no signed consent or documentation of compensation terms.
- Baseline Social & Environmental Studies
    - Audit Check: Confirm that baseline socio-economic and environmental studies (household census, livelihood mapping, water-use survey, cultural/forest-rights mapping) were completed and approved before large-scale exploration.
    - Risk: Future R&R cost blow-outs and legal disputes if baseline conditions are undocumented or manipulated.
- Stakeholder Engagement & Grievance Records
    - Audit Check: Review minutes of public meetings/Gram Sabha resolutions, grievance registers, and commitment logs (promised jobs, roads, water schemes) maintained during exploration.
    - Fraud Indicator: CSR/R&R promises made in field meetings but absent from any formal register or budget.

3.4. Detailed Audit Work Program

Sub Process	Red Flag / Risk	Audit Procedure (Field Steps)
1. License & Rights Management	Illegal Exploration: Drilling outside the designated "Prospecting License" (PL) area, leading to fines or cancellation.	Coordinate Overlay: Obtain the GPS coordinates of the current drilling sites. Overlay them on the digitized "PL Boundary Map." Ensure no rig is operating in a "No-Go Zone" (Forest/Revenue land without permit).
1. License & Rights Management	Expired Rights: Continuing exploration activities after the PL validity has expired	Validity Check: Review the PL Deed. Compare the "Expiry Date" with the dates on the "Daily Drilling Report." Any drilling post-expiry is illegal expenditure.

	without a formal renewal application.	
2. Drilling Operations (Physical Verification)	Ghost Meters: Contractor billing for 100 meters of drilling when only 80 meters were actually drilled.	The "Stick-Up" Measurement: Visit the rig. Measure the drill rods currently in the ground + the rods stacked on the truck. Compare this total length with the depth claimed in the "Shift Report."
2. Drilling Operations (Physical Verification)	Poor Core Recovery: Low recovery (<90%) in the mineralized zone, leading to unreliable grade estimation.	Core Run Check: Go to the core shed. Pick a 3-meter run. Measure the actual rock inside. If it is only 2 meters long, Recovery = 66%. Flag this as "Unreliable Sample."
3. Sampling & Quality Assurance	Salting (Fraud): Deliberately adding high-grade concentrate or coins to the sample bag to inflate the assay results.	Blank Sample Test: Check the assay results for "Blank Samples" (sterile sand/rock inserted blindly). If the Blank returns a high metal grade, the batch is contaminated or salted.
3. Sampling & Quality Assurance	Bias in Core Splitting: The technician selects the "good looking" half of the core for the lab and leaves the waste half in the box.	Visual Comparison: Inspect the "Archive Half" remaining in the core box. Does it look visually poorer (less mineralization) than the assay results of the "Lab Half"? If yes, splitting was biased.
4. Resource Estimation (The G-Axis)	Category Inflation: Classifying "Inferred Resources" (G3) as "Proved Reserves" (G1)	Drill Grid Verification: Measure the distance between boreholes on the map. If they are >200m apart, the area is G3. Challenge

	to boost valuation or bankability.	the Geologist if they have marked it as G1 (which requires <50-100m spacing).
4. Resource Estimation (The G-Axis)	Extrapolation Risk: Assuming the ore body continues far beyond the last drill hole.	Polygon Check: In the Reserve Estimation software/map, check the "Influence Polygon" around the last drill hole. It should not extend more than 50% of the grid spacing.
5. Core Library Management	Destruction of Evidence: Core boxes are left in the open rain, rotting or mixing, making re-verification impossible.	Physical Inventory: Select 5 random boreholes from 2 years ago. Ask to see their boxes. If the labels are faded or boxes are missing, the "Geological Asset" is unverifiable.
6. Exploration Accounting	Capitalization of "Dry Holes": Capitalizing costs of drilling that found no minerals, inflating the Asset Book.	Success-Efforts Test: Trace the cost of specific boreholes. If a borehole report says "Barren" or "No Significant Mineralization," its cost must be Expensed to P&L, not Capitalized as CWIP.
7. Contractor Management	Fuel Theft (Drilling): Rig operators siphoning diesel provided by the mine.	Consumption Norm: Calculate Litres per Meter Drilled. Compare it with the rig manufacturer's spec. A 20% variance indicates theft or extreme engine inefficiency.

3.5. Sector Specifics: The Commodity Nuances

Feature	Coal Exploration	Metal Mining (Iron/Zinc)
Drilling Type	Often Non-Coring (Open Hole) for upper layers to save cost. Only coal seams are cored.	Diamond Coring is standard to understand structure and grade.
Key Metric	Seam Thickness. (How thick is the coal layer?)	Grade Continuity. (Does the vein continue?)
Geophysical Logging	Mandatory. Gamma-ray logging is used to identify coal vs. shale inside the hole.	Used less frequently; chemical assay is primary.
Audit Focus	Audit the Geophysical Logs (digital readout) vs. the Manual Log.	Audit the Assay Lab results.

3.6. Common Findings & Reporting

- **Finding:** "Drilling Rigs Idle Time" charged to the company.

Observation: The contract allows standby charges only if the *Company* causes the delay (e.g., site not ready). However, the contractor billed standby for their own breakdowns.

- **Finding:** No "Core Recovery" penalty deducted.

Observation: The contract states: "If core recovery is <90%, no payment." The log showed 60% recovery, but full payment was released.

4.1 Introduction & Audit Objective

Between finding the ore (Exploration) and digging it out (Production), lies the critical phase of Mine Planning and Project Construction. This is where the "Feasibility Study" is converted into steel and concrete.

For the Internal Auditor, this constitutes the highest financial risk phase. Mining projects are notorious for cost overruns and schedule delays. The objective of this chapter is to audit the "Build Phase"—ensuring that the mine is designed economically, the Capex budget is realistic, and the project engineering contracts protect the company's interest.

Primary Audit Risks:

- "Gold Plating": Engineering designs that are excessively expensive or over-specified for the actual need.
- Capex Blowout: Inaccurate initial budgeting (Project Cost Estimation) leading to fund shortages.
- Scope Creep: Uncontrolled changes to the project scope after the budget is approved.
- EPC/EPCM Leakage: Poor management of Engineering, Procurement, and Construction contracts.

4.2 Regulatory Reporting Standards. This is crucial because financial auditors often rely on geological reports to value the "Intangible Assets" or "Mining Rights" on the Balance Sheet.

The "Truth Serums": JORC, NI 43-101, & The Indian Code

Geology is an estimation, not an exact science. To prevent companies from lying to investors (telling them "We found Gold" when they only found "Yellow Dirt"), the world established strict Reporting Codes. These codes dictate HOW a company can publicly claim it has "Reserves." If a report isn't signed off according to these codes, the numbers are legally worthless.

1. The Global "Big Three"

Standard	Origin	The Auditor's Take
JORC Code	Australia	The Commercial Standard. Most common globally. If a bank or private equity firm is funding the mine, they will demand a JORC report. It relies heavily on the "Competent Person" (CP) taking responsibility.
NI 43-101	Canada	The Strict Legal Standard. Born after the massive "Bre-X Gold Fraud." It is a *law*, not just a guideline. It requires specific technical formats. If a company is listed on the TSX (Toronto), this is mandatory.
SAMREC	South Africa	The Deep Mining Standard. Similar to JORC but tailored for complex deep-level mining common in South Africa (Gold/Diamonds).

The Common Thread (CRIRSCO): All these codes (and the Indian IMIC) fall under the CRIRSCO umbrella (Committee for Mineral Reserves International Reporting Standards). They all speak the same language:

- You cannot call it a "Reserve" (Bankable) unless you have done a Feasibility Study.
- The report must be signed by a Competent Person (CP) / Qualified Person (QP).

2. The Indian Equivalent: A "Dual System"

India is unique. As an auditor, you will encounter two different standards depending on who the report is for.

A. The Statutory Standard (For Government/IBM): UNFC

- Name: United Nations Framework Classification
- Mandate: The Indian Bureau of Mines (IBM) legally mandates this for all Mining Plans and Returns.

- The Structure: It uses a 3-digit code (Axis).
  - E (Economic Viability)
  - F (Feasibility Study)
  - G (Geological Knowledge)

  Example: 111 is a Proven Reserve. 333 is an Inferred Resource.
- Audit Relevance: You check this for Royalty and Statutory Compliance.

B. The Investor Standard (For IPOs/Banks): IMIC

- Name: Indian Mineral Industry Code (IMIC) / The NACRI Code.[3]
- Status: It is the Indian member of CRIRSCO (equivalent to JORC).[4]
- Mandate: Adopted by SEBI for listed companies.
- Audit Relevance: You check this for Asset Valuation, Impairment Testing, and Bank Loans.

3. The Audit Checklist: "Who Signed It?"

The most critical control in all these codes is the Competent Person (CP).

Audit Checkpoint	The Procedure	The Red Flag
1. The Signatory	Check the "Resource Report." Who signed it?	Self-Signed: If the Mine Manager signed it, it's invalid. It *must* be signed by a registered "RQP" (Recognized Qualified Person) or a NACRI member.
2. The Conflict	Is the Competent Person an employee of the company?	Internal Valuation: While allowed in some codes, for high-value assets, an *Independent* CP (External Consultant) is required to avoid bias.
3. The Basis	Does the report reference "UNFC	Mixed Signals: Using "UNFC 333" (Inferred) to justify a Bank Loan.

	111" or "Proved Reserves"?	Banks usually fund only 111 or 121 categories.

Auditor's Rule of Thumb:

- If checking Government Dues > Follow UNFC (IBM).
- If checking Balance Sheet Value > Ask for IMIC / JORC.

4.3 Key Audit Focus Areas

A. The Feasibility Study (The Bible)

Every mine is built based on a Techno-Economic Feasibility Report (TEFR) or a Bankable Feasibility Study (BFS). The auditor must benchmark all subsequent spending against this document.

The Audit Check: Does the actual equipment ordered match the capacity specified in the TEFR? (e.g., If the TEFR specifies 100-tonne dumpers, why are we buying/renting 240-tonne dumpers?)

B. Project Cost Estimation (AACE Classes)

Auditors must understand the accuracy of the budget. Mining projects use the AACE International Cost Estimate Classification System. (not a mandatory basis but acceptable commercial standard used widely as Industry Practice).

Risk: Management treating a "Class 4" estimate (Concept, +/- 30% accuracy) as a "Class 1" (Control, +/- 10% accuracy), leading to inevitable budget variance.

AACE Class	Project Stage	Accuracy Range	Who Performs It? (The Owner)	Audit Check / Methodology
Class 5	Concept / Scoping	-50% to +100%	Strategic Planning / Geology Team *(Often internal staff using*	Benchmarking: Check if they used historical data from similar mines

			"Rules of Thumb")	
Class 4	Pre-Feasibility (PFS)	-30% to +50%	Mine Planning Engineer *(Internal team + Preliminary vendor quotes)*	Factoring: Audit the "Equipment Factors." Did they budget the Crusher cost based on a vendor email or just a website price?
Class 3	Feasibility (FS)	-20% to +30%	EPCM Consultant (e.g., Bechtel, Jacobs, Hatch) *(Qualified Cost Engineers)*	The "Bankable" Test: This is the budget submitted to the Board/Bank. Verify that >40% of the engineering design is complete.
Class 2	Control / Bid	-15% to +20%	Procurement Head / EPC Contractor *(Based on actual tenders received)*	Quote Verification: Trace the "Steel Structure" cost to the L1 Bidder's final offer, not the initial estimate.
Class 1	Execution / Check	-10% to +15%	Project Controls Manager / QS *(Based on "Issued for Construction" drawings)*	Change Order Audit: If the actual cost deviates from Class 1, it indicates execution failure, not estimation error.

C. Engineering Contracts: EPC vs. EPCM

Understanding the contract type is vital for auditing payments.

- EPC (Engineering, Procurement, Construction): A "Lump Sum Turnkey" contract. The contractor takes the risk.

  Audit Focus: Ensure no "Change Orders" are approved for items already in the fixed scope.

- EPCM (Engineering, Procurement, Construction Management): The contractor acts as a consultant/manager; the mine owner pays the direct costs.

  Audit Focus: Verify vendor invoice and the manager's time-sheets.

D. Project Monitoring (The S-Curve)

The "S-Curve" plots cumulative costs or progress against time.

- The "Banana" Curve: Auditors should look at the gap between "Early Start" and "Late Start" curves. If the actual progress drops below the "Late Start" curve, the project is officially delayed, and penalties (Liquidated Damages) should be triggered.

E. Life-of-Mine Plan & Reserve Governance

The Life-of-Mine (LoM) plan is the financial backbone of the project; every NPV, impairment test and fund-raise rests on it.

Sub Process	Risk	Audit Procedure
Competent Person & Reserve Sign-off	Over-optimistic reserve claims used for bank finance or IPOs while statutory filings show lower figures.	Audit Check: Verify that Ore Reserves and Mineral Resources are signed off by a "Competent Person" (as per JORC/NI 43-101/SAMREC/Indian equivalent) and that the same figures flow into the TEFR/BFS, mining plan and board presentations.

LoM Updates & Impairment Linkage	Red Flag: Large downgrades of reserves in internal technical reports with no corresponding impairment or disclosure.	Audit Check: Review how often the LoM plan is updated (e.g., annually) and whether updated strip ratios, grades and prices are fed into Ind AS 36 impairment tests for the mine CGU.
Scenario & Sensitivity Analysis	Operating on out-of-date, “bull-market” assumptions.	Audit Check: Inspect sensitivity cases (price, grade, diesel cost, SR) used for investment approval and check if actual conditions have already breached “downside” cases without revisiting the plan.

4.4 Detailed Audit Work Program (Checklist)

Sub Process	Red Flag / Risk	Audit Procedure (Field Steps)
1. Budgeting & Estimation	The "Class Jump" (Budget Variance): Management treating a "Class 4" estimate (Concept, +/- 30% accuracy) as a "Class 1" budget, leading to inevitable cost overruns.	Basis of Estimate (BOE) Review: Review the Capex budget supporting. Are costs based on firm vendor quotes (Class 2) or historical "Rules of Thumb" (Class 4/5)? Flag any sanction request lacking Class 2 maturity.

2. Contingency Management	Misuse of Reserves: The "Contingency Fund" (usually 5-10%) being used to cover up poor planning or scope creep rather than genuine unforeseen events.	Utilization Audit: Scrutinize every draw-down from the Contingency Budget. Was it for a "Risk Event" (e.g., flood) or a "Planning Error" (forgot to order cables)? Report planning errors separately.
3. Land & Site Readiness	Idle Time Claims: Awarding Engineering contracts (EPC) before 100% land possession is secured, leading to massive "Mobilization Advance" payments for idle contractors.	Right of Way (ROW) Check: Verify "Encumbrance Free" land status *before* approving the release of Mobilization Advances. Ensure the site is physically ready for the contractor to start work immediately.
4. Engineering Design	"Gold Plating": Over-specifying equipment (e.g., using stainless steel piping where carbon steel is sufficient) to inflate project cost, often driven by conservative engineers or vendor influence.	Spec Comparison: Compare the design specifications with the Feasibility Study (TEFR) and industry standards. Challenge any "Specification Upgrade" that does not have a clear ROI calculation.

5. Procurement & Tendering	Vendor Favouritism: Manipulating the "Technical Evaluation Criteria" (high weightage to non-essential features) to favor a specific expensive vendor over a cheaper, capable one.	Scorecard Audit: Review the Tender Evaluation Sheet. Look for subjective criteria (e.g., "Vendor Reputation") that were used to disqualify the L1 (Lowest) bidder.
5. Procurement & Tendering	Structuring (Split POs): Splitting a large Capex order (e.g., 1 crore) into five smaller POs (20 lakhs each) to bypass higher-level scrutiny / evaluation.	Vendor Concentration Check: Filter the PO dump for the same vendor within a 30-day window. Consolidate the values. If the total exceeds the approval limit of the signer, it is a governance breach.
6. Contract Execution	"Change Order" Tsunami: A contractor wins the bid with a low price but later submits dozens "Scope Variations" to increase the price post-award.	Variation Order Analysis: Calculate the ratio of Total Variation Value / Original Contract Value. If > 10%, audit the "Scope Definition" in the original tender for gaps.

6. Contract Execution	Front-Loading: The contractor alters the "Payment Schedule" to collect 80% of the money while completing only 40% of the work, leaving the mine owner at risk.	Milestone Verification: Compare "Cumulative Payments %" vs. "Physical Progress %" (S-Curve). If Financial Progress > Physical Progress, stop payments immediately.
7. Civil Works Verification	Billing for Air: Paying for "Running Account" (RA) bills for civil works (excavation/concrete) that were never executed or are over-measured.	Measurement Book (MB) Reconciliation: Select a sample of certified RA bills. Physically measure the dimensions on-site or cross-check with "Issued for Construction" drawings. Verify if the "Hidden Items" (foundations) match the photos taken during pouring.
8. Material Management	Theft of Free-Issue Material: Contractors stealing cement/steel provided by the company ("Free Issue") and claiming it was consumed in the work.	Reconciliation Audit: Calculate Theoretical Consumption (Drawings) + Allowable Wastage (2%). Compare this with Actual Issued Quantity. Any excess issue must be recovered from the contractor's bill.
9. Geotechnical Safety	Production Overriding Safety: Operational teams steepening slope angles or thinning pillars to meet production targets, ignoring the design parameters.	Design Compliance Check: Review "Design Change Notes." Did the Geotechnical Engineer sign off? If a production manager approved a slope change without geotechnical sign-off, it is a major safety violation.

10. Design Document	Production teams steepening slopes or thinning pillars to meet short-term targets	Obtain geotechnical design reports (bench heights/widths, overall slope angles, berms, underground support patterns) and verify that the operating pit/stopes match the design sections.
11. Monitoring & Trigger Action Response Plans (TARPs)	Instruments installed but readings not logged or no documented action when thresholds were breached	Review prism/inclinometer records, convergence monitoring, and TARPs specifying when to evacuate areas or stop mining.
12. Geo Technical Governance	"Production-only" approvals on design change notes where geotechnical sign-off is missing.	Confirm that geotechnical engineers sign off on design changes, pushbacks and dump raisings, not just mine production managers.

4.5 Case Study: The "Gold-Plated" Crusher

Brief Case Study: A mining company ordered a primary crusher with a capacity of 5,000 TPH (Tonnes Per Hour) based on an optimistic "Expansion Phase" that was 10 years away. The current mine output was only 1,500 TPH. Result: Massive capital lock-up, higher maintenance costs for an oversized engine, and depreciation hitting the P&L for an asset utilized at only 30% capacity.

Part II – Core Operations

Chapter 5: Audit of Mine Planning & Pre-Stripping

5.1 Introduction & Audit Objective

Before a single ounce of ore is extracted, the mine must be "opened up." This phase involves Mine Planning (designing the pit) and Pre-Stripping (removing the waste/overburden to expose the ore).

For the Internal Auditor, this is a high-stakes chapter. Pre-stripping often consumes 40-60% of the initial working capital. If the planning is flawed, or if the "Stripping Ratio" is manipulated, the mine's entire financial model may fail.

Primary Audit Risks:

- Optimistic Planning: Mine plans that assume unrealistic equipment availability or geological grades.
- Capitalization Fraud: Treating operational waste removal as "Pre-Stripping" (Capex) to inflate immediate profits.
- The "Lead Distance" Trap: Contractors dumping waste closer than planned to save fuel, filling up areas needed for future mining.
- Stripping Ratio (SR) Manipulation: Hiding waste costs to lower the reported "Cost of Production."

5.2 Key Audit Focus Areas

A. The Mine Plan & The "Block Model"

The "Block Model" is the digital 3D representation of the ore body. The Audit must verify if the Short Term Mine Plan (STMP) aligns with the Life of Mine (LOM) plan.

- Risk: Operations team mining "easy" high-grade ore now (High-Grading) to meet annual targets, leaving difficult/expensive areas for later years, destroying long-term value.
- Check: Compare the *planned* stripping ratio vs. the *actual* stripping ratio for the month.

B. Pre-Stripping: Capex vs. Opex (The IFRIC 20 Standard)

This is the most technical accounting area in mining.

- Pre-Stripping (Development Phase): Removal of OB to *initially expose* the ore body. This is Capitalized (Asset).
- Production Stripping (Production Phase): Ongoing removal of waste to maintain access. This is Expensed (Cost).
- Audit Insight: Auditors must check if the company is capitalizing *ongoing* stripping costs to artificially boost profit margins.

C. Stripping Ratio (SR) Verification

Stripping Ratio is the volume of waste removed to extract one unit of ore.

Audit Check: If the SR decreases suddenly without geological justification, the mine might be "under-stripping" (falling behind on waste removal), which is a massive future liability.

5.3 Detailed Audit Work Program (Checklist)

Sub Process	Red Flag / Risk	Audit Procedure (Field Steps)
1. Stripping Ratio (SR) Compliance	High-Grading (Short-Termism): The mine removes 1 ton of Ore but only 2 tons of Waste, against a planned SR of 1:4. This boosts current profit but creates a massive "Waste Wall" liability for the future.	Cumulative SR Check: Don't just check the monthly SR. Calculate the "Year-to-Date (YTD) Stripping Ratio." If YTD SR < Planned SR by more than 10%, report it as a "Governance Risk".
2. Geological Reconciliation	Model Failure: The actual grade recovered is consistently lower than the Block Model prediction, leading to	The "F1 Factor" Check: Ask for the monthly reconciliation report. Compare Actual Tonnage & Grade vs. Block Model Tonnage & Grade. If variance >

	"Write-offs" of reserves later.	10%, investigate if the model is over-optimistic.
3. Baseline Survey Control	The "Paper Hill" Fraud: Contractors billing for removing a hill that didn't exist by manipulating the initial ground level (OGL) survey.	OGL Sign-off: Verify the "Original Ground Level" (OGL) document. Was it jointly signed by the Surveyor, Contractor, and Mine Manager *before* the first scoop of earth was removed?. Physical verification can also be done.
4. Contractor Billing (Volume)	The "Swell Factor" Arbitrage: Contractors billing based on "Truck Counts" (Loose Volume) instead of "In-Situ Survey" (Bank Volume). Soil swells by 20-30% when dug. Paying on loose volume overpays by 30%.	Survey Reconciliation: Verify that the "Abstract of Bill" (RA Bill) is supported by a Joint Survey Report (Initial Level- Final Level). If billing is on "Trip Count," apply the standard "Swell Factor Correction" (e.g., Divide quantity by 1.3).
5. Lead Distance (Transport)	Lead Inflation: The contractor claims the dump yard is 4 km away (Lead Slab 3-4km), but a shortcut exists making it 2.5 km (Lead Slab 2-3km).	Odometer Test: Physically ride in a dumper from the Pit Face to the Dump Yard. Note the odometer reading. Compare this "Actual Lead" with the "Weighted Average Lead" claimed in the monthly bill.
6. Waste Dump Management	Sterilization of Reserves: Dumping waste rock on top of land that contains	Coordinate Overlay: Ask for the "Dump Coordinates." Overlay them on the "Geological Resource Map" (G2/G3 layer). If the dump sits on a mineralized

	future ore reserves underneath.	zone, raise a Major Non-Compliance.
7. Pre-Stripping Capitalization	Opex as Capex: Capitalizing routine waste removal as "Development Cost" (Capex) to artificially inflate the company's P&L profits.	The "Touch Ore" Test: Verify the date when the mine first "touched" the ore body. Any waste removal *after* sustainable production starts must be Expensed (Opex), not capitalized.
8. WIP Inventory Valuation	Inventory Inflation: Valuing "Pre-stripped Ore" (ore exposed but not extracted) at the "Final Product" cost, inflating current assets.	Valuation Test: Review the WIP valuation sheet. Ensure pre-stripped ore is valued only at the *direct cost of stripping*, excluding processing or overhead costs.
9. Mine Design Compliance	Slope Failure Risk: Making the "Bench Height" too high or "Bench Width" too narrow to extract more ore, risking a landslide.	DGMS Rule Check: Measure the bench height with a laser rangefinder. If the permission says "10 meters" but the actual wall is 15 meters high, it is a statutory violation (Section 106 of MMR).

5.4 Red Flags & Fraud Indicators

- "Short-Dumping": Satellite imagery or drone surveys showing waste dumps encroaching on the mine pit boundary (contractors dumping too close).
- Sudden Drop in SR: The mine reports record low costs because they stopped removing waste (SR drops), but this creates a "wall of waste" that will kill production next year.

- Survey Discrepancies: The contractor's claimed volume (truck counts) is consistently 15-20% higher than the drone/Total Station survey volume.
- The "Paper Hills" – *Contractor billing for removing a hill that didn't exist (manipulating the Initial OGL survey), resulting in fraudulent payouts.*

5.5 Technical Brief: Bulking Factor (Swell Factor) & Volume Reconciliation

This sub chapter will help equip the auditor with the math and methods to challenge contractor invoices.

1. The Core Concept: The "Three States" of Earth In mining contracts, ambiguity about which "state" of volume is being measured is the #1 cause of overpayment. Material exists in three distinct densities:

  - Bank Volume (BCM- Bank Cubic Meter): The natural, *in-situ* state (solid rock/earth) before excavation. This is the hardest to measure but the only fair basis for payment.
  - Loose Volume (LCM - Loose Cubic Meter): The material after blasting/digging. It contains air voids. This is what trucks carry.
  - Compacted Volume (CCM- Compacted Cubic Meter): The material after it has been dumped and rolled/pressed (e.g., in a waste dump or road).

The Trap: Contractors prefer to get paid on LCM (Truck Counts) because it is 20-40% higher than BCM. Mine owners must insist on paying on BCM (Survey Difference).

2. The Formulas

  - Swell %: The percentage increase in volume when rock is broken.
  - Swell Factor: The multiplier used to convert Bank Volume to Loose Volume.
  - Load Factor : The inverse of the Swell Factor. This is used to convert "Truck Volume" back to "Bank Volume" for reconciliation.

3. Standard Swell Factors (Reference Table) *Note: These are industry averages. The auditor must demand site-specific lab tests.*

Material	Swell %	Swell Factor	Load Factor
Sand (Wet)	10- 15%	1.10- 1.15	0.87- 0.90
Top Soil / Clay	20- 40%	1.20- 1.40	0.71- 0.83
Coal (Blasted)	30- 35%	1.30- 1.35	0.74- 0.76
Limestone	50- 65%	1.50- 1.65	0.60- 0.66
Hard Rock (Granite)	60- 75%	1.60- 1.75	0.57- 0.62

Audit Rule of Thumb: Harder rock = Higher Swell. If a contractor claims a 20% swell for Granite, they are likely under-reporting the density to hide poor truck utilization.

4. The "Million Dollar" Audit Calculation (Example)

Scenario: A contractor is removing Overburden (Hard Sandstone).

Contractor Claim: "We transported 10,000 truck trips. Our dumpers have a volumetric capacity of 20m3 Therefore, we excavated 200,000 m3".

Contract Rate: 400 per m3. Invoice Value: 8,00,00,000.

The Auditor's Challenge:

The contractor is billing based on Loose Volume (LCM) capacity of the truck. The contract likely specifies payment on Bank Volume (BCM).

1. Determine Material Properties: Hard Sandstone has a Swell Factor of approx 1.50 (50% swell).
2. Calculate Load Factor: LF = 1 / 1.50 = 0.666
3. Convert to Bank Volume:

Actual Excavated BCM = Total Truck Volume X Load Factor

200,000 X 0.666 = 133,333

4. Correct Valuation: 133,333 x 400 = 5,33,33,200

Audit Finding: The contractor attempted to overcharge by (33%) by ignoring the Bulking Factor.

5. Field Audit Procedure: The "Box Test"

How does an auditor verify the Swell Factor if the contractor disputes the theoretical table?

You perform a "Box Test" on site.

Step 1: Build the Box

Construct a sturdy wooden or steel box with internal dimensions exactly 1m x 1m x 1m (Volume = 1 m^3).

Step 2: Weigh the Bank (In-Situ)

- Core drill or cut a precise 1 m^3 block of the rock *before* blasting (difficult) OR (easier) rely on the specific gravity (SG) from the geologist's lab report (e.g., SG of Sandstone = 2.4 tonnes/m^3).

Step 3: Weigh the Loose

- Fill the 1 m^3 box with the *blasted/excavated* rock.
- Weigh the box (Gross Weight- Empty Box Weight = Net Weight of Loose Rock).
- *Example:* Net Weight = 1.6 tonnes.

Step 4: Calculate Swell Factor=Density Bank/Density Loose = 2.4/1.6 = 1.5

6. Recommended Contract Clauses (Protective Wording)

Clause xx.xx- Measurement of Quantities: "All payments for excavation shall be based on Bank Cubic Meters (BCM) as determined by joint survey (Initial OGL vs. Final Level) using Total Station/Drone photogrammetry. Truck counts are for operational estimation only and shall not be used for billing."

Clause xx.xx- Bulking Factor Application: "Where measurement by survey is not possible and truck counts must be used, a fixed Load Factor of 0.XX (or Swell Factor of 1.XX) shall be applied to convert Loose Volume to Bank Volume. This factor shall be determined by a joint field test conducted once every 6 months.

Chapter 6: Production & Extraction

Chapter 6 is the heartbeat of the mine. This is where the daily operational expenditure (Opex) is incurred. In the context of Indian and Commonwealth mining contracts, "Raising" typically refers to the extraction and bringing of mineral to the surface (often outsourced as "Raising Contracts").

6.1 Introduction & Audit Objective

While Chapter 5 covered moving waste (Pre-stripping), Chapter 6 covers the "Revenue Cycle": getting the ore out. The production cycle follows a rigid sequence:

The Value Chain: Drill Planning > Blasting > Excavation (Loading) > Hauling > Crushing >Stockpiling

For the Internal Auditor, the objective is to verify Efficiency (Cost per Tonne) and Quantity (Prevention of theft/leakage). This phase accounts for 60-70% of the mine's recurring cash outflow.

Primary Audit Risks:

- Poor Fragmentation: Bad blasting creating "boulders" that slow down excavators and damage crushers (Hidden Cost).
- Explosive Handling & Pilferage: High value, high risk. High accident impact on misuse. Diversion of explosives to illegal markets.
- Raising Contract Leakage: Overpaying contractors for "re-handling" or claimed distances that don't exist.
- Diesel Theft: The most common fraud in the hauling phase.

6.2 Key Audit Focus Areas

A. Drilling & Blasting (The Cost Driver)

Blasting is the first step in comminution. If rock isn't broken right here, it costs 3X more to break it later in the crusher.

- Powder Factor (PF): The critical efficiency metric.

Powder Factor (PF) = Total Explosives Used (kg) / Total Rock Broken (Tonnes)

- Audit Check: Compare *Planned PF* vs. *Actual PF*.
  - If Actual PF is High: Potential theft of explosives or incompetent blasting design.
  - If Actual PF is Low: Be careful. This might look like "cost saving," but it usually results in "Oversize" boulders that kill excavator productivity.
- Typical range:
  - Solid blasting: 1.5 – 2.5 tonnes/kg
  - Cut blasting: 4 – 5 tonnes/kg

B. The "Raising" (Excavation & Hauling)

This refers to the lifting of ore from the pit bottom to the surface stockyard.

- Match Factor: The ratio of Truck capacity to Shovel capacity. In mining operations, Match Factor (MF) is a measure of how well the capacities of excavators (shovels) and dumpers (trucks) are balanced. It is defined as the ratio of the time required by a dumper to complete a cycle to the time required by the excavator to load that dumper.
- MF=Truck Cycle Time / Shovel Loading Time

  Truck Cycle Time (TCT): Time taken by a dumper to travel, dump, return, and queue.

  Shovel Loading Time (SLT): Time taken by the excavator to load one truck fully.
- Audit Insight: A Match Factor of 1.0 is perfect. Match Factor is a synchronization metric in mining logistics. A value close to 1 means optimal balance between shovel and dumper, while deviations highlight inefficiencies in fleet management. A perfect number may also reflect manipulation in data.
  - < 1.0: Shovels are waiting (Under-trucked). Production loss.

  - > 1.0: Trucks are queuing (Over-trucked). Fuel waste.

C. Contractor Management (Raising Contracts)

Many mines outsource this entire cycle.

- Escalation Clauses: Audit the "Price Variation Clause" (PVC) for Diesel. When diesel prices drop, does the contractor's rate per tonne drop immediately?
- Re-handling: Verify if the contractor is double-billing for moving the same ore twice

  (e.g., Pit > Temporary Stockpile > Crusher).

D. Operational Technology (OT) Data Integrity

Most production numbers now originate from Fleet Management Systems (FMS), GPS and automation, not from clipboards.

- System Access & Overrides
  - Audit Check: Review user access and change logs for FMS, dispatch systems and drill/loader telemetry. Ensure supervisors cannot retro-edit trip counts, cycle times or locations without dual approval.
  - Red Flag: Night-shift supervisors with “admin” rights and no independent review of changes.
- Reconciliation with Physical Reality
  - Audit Check: Reconcile FMS production by face/shift with weighbridge totals, diesel issues and blasting tonnage. Spot faces that show high tonnage with low diesel and explosives consumption.
  - Risk: “Virtual tonnes” created in the FMS to inflate contractor bills or internal KPIs.

6.3 Detailed Audit Work Program (Checklist)

Sub Process	Red Flag / Risk	Audit Procedure (Field Steps)
1. Drilling Operations (Contractor billing)	Phantom Drilling: Contractor billing for "Air Drilling" (holes that don't exist) or over-reporting depth to inflate the invoice.	Drill Log Verification: For a sample physical drill holes- drop a weighted tape measure to verify actual depth. Compare with the "Drill Log" signed by the supervisor. If the log says 10m but tape hits bottom at 8m, it is fraud.
2. Explosives inventory	Theft of Sensitive Material: Detonators and Boosters are high-value/high-risk items often diverted to illegal markets.	Magazine Reconciliation: Perform a surprise "Floor-to-Book" count. Any variance is a critical security risk.
3. Explosives misfire management	Explosives that "didn't go off" are claimed to be destroyed but are actually stolen.	Destruction Protocol: Review the "Misfire Register." If a misfire occurred, trace it to the "Destruction Memo" countersigned by the Safety Officer. If there is no memo, the explosive is missing.
4. Explosives Blasting efficiency	Inflated Production: The mine claims to have blasted 50,000 tonnes, but the "Heave" (broken rock pile) looks significantly smaller.	Post-Blast Survey: Whether the surveyor measured the volume of the "Heave" immediately after the blast using a drone/Total Station. Or the reports were made after wards which could have been just signed off. There are various scientific methods*

		adopted by surveyors to accomplish this.
5. Dumper Loading efficiency	Under-Loading: Dumpers meant to carry 100T are carrying only 85T, increasing the "Cost per Tonne" significantly.	Bucket Fill Factor: Physically observe loading for 1 hour. Are excavators filling dumpers to the brim? Calculate average payload (Total Weighbridge Wt / Trip Count) vs. OEM rated capacity.
6. Dumper Haulage Verification	Phantom Trips: Trucks creating fake trips in the logbook (or FMS) without actually traveling to the pit.	Trip Sheet vs. GPS: For random trips, cross-verify with GPS playback. Did the truck actually enter the "Ore Plot" geofence?
7. Dumper Cycle Time Analysis	Over-Trucking (Inefficiency): Paying for rental trucks that spend half the shift waiting in queue (idling) at the excavator.	Match Factor Check: Check for idling and verify "Queuing Time" in Fleet Management System (FMS) reports.
8. Dumper-FMS Data Integrity	"Virtual Tonnes": Supervisors with "Admin" access to the Fleet Management System retro-actively editing trip counts to inflate bills.	Audit Log Review: Extract the "User Activity Log" from the FMS. Filter for commands like "Edit Trip" or "Insert Trip". Frequent edits during Night Shift without justification indicate data fraud.
9. Dumper-Dispatch Control	Illegal Mining: Trucks leaving the mine with ore without being	Weighbridge Bypass: Inspect the pit exit route. Are there "Bypass Roads" or broken barriers allowing trucks to leave without

	weighed, facilitating off-book sales.	weighing? Verify CCTV footage for unauthorized exits.
10. Dumper – Efficiency	Residual sludge/ROM/ore not fully offloaded; dumpers continue hauling with partial loads, leading to inflated trip counts and misrepresented production.	Conduct random physical inspections of dumper beds post-unloading. Review CCTV footage at loading/unloading points.
11. Diesel Escalation	Financial Overpayment: Paying higher rates due to diesel hikes even when base mileage assumptions are incorrect.	Escalation Formula Audit: Re-calculate the price adjustment. Ensure "Base Price" matches the *original tender date* (not current) and "Fuel Norm" (Litres/Tonne) is fixed.
12. Re-handling Check	Double Billing: Contractor billing for moving the same pile twice (Pit to Stockpile to Crusher) as two primary trips.	Mass Balance Check: Compare Tonnage Mined (Pit) vs. Tonnage Fed (Crusher). If the contractor claims to have moved 120 tonnes for every 100 tonnes produced, investigate the loop.
13. Grade Control	Quality Dilution **: Contractor mixing Waste with Ore to increase tonnage, destroying the "Feed Grade".	Penalty Clause Check: Review assay reports. If grade < Cut-off due to dilution, verify if the contractual penalty was deducted from the bill. *(Detailed steps given after the table)*
14. De-Watering Ops	Ghost Running Hours: Billing for pumps	Specific Power Check: Calculate kWh Consumed / (Flow Rate x Head). If pumps consume power

	running 24/7 during dry months.	but shows low water discharge, they are idling.
15. Secondary Breakage	Inefficiency Shifting: Low Powder Factor (cheap blasting) leads to boulders, forcing expensive "Rock Breaker" hiring.	Rock Breaker Utilization: Compare "Breaker Hours" vs. "Powder Factor." If PF drops while Breaker Hours rise, the mine is inefficiently shifting cost to machinery hire.
16. Explosives Logistics	Detention Charges: Vans kept waiting at the gate for days due to full magazines or absent staff.	Van Detention Audit: Check 'Gate Entry' vs. 'Magazine Unloading' time. If vans wait >4 hours, investigate planning failures.

*** Methods for blasting efficiency (Refer Checkpoint 4 above) measurement by surveyor:**

- Method 1: The "Digital Twin" Volumetric Reconstruction
- Method 2: Excavator "Dig Time" Forensics (The Physics Check)
- Method 3: Image Granulometry (WipFrag Analysis)

Auditor's Note: If the auditee says, *"We didn't fly a drone and we have no photos,"* record this as a Major Internal Control Failure. Visual inspection of a blast face is unsafe (risk of loose rocks/misfires); remote inspection (drone) is the industry safety standard. Missing data is a red flag in itself.

**** Quality Dilution by Contractor (Refer Checkpoint 13 above):** This is one of the most common conflicts in mining: The Contractor is paid for Quantity (Tonnes), but the Mine Owner makes money on Quality (Grade). The contractor has a financial incentive to dig up "dirt" along with the ore to make the trucks heavier and bill more.

Here is the detailed breakdown of at which stages this mixing happens and how to catch it.

The Three Stages of "Dilution" (Mixing Waste into Ore)

Stage 1: The Floor (Vertical Dilution)- *Most Common*

- The Scenario: The ore body is a 5-meter thick layer. Below it is waste rock or mud.
- The Contractor's Trick: Instead of stopping exactly at 5 meters, the excavator operator digs 5.5 meters deep.
- The Result: That extra 0.5 meters is pure waste (heavy rock/mud). It adds 10% more weight to the truck (Contractor gets paid more), but it lowers the Ore Grade from 62% Fe to 58% Fe (Owner loses revenue).
- Audit Term: "Digging the Toe" or "Floor Dilution."

Stage 2: The Wall (Horizontal Dilution)

- The Scenario: The ore body is a vertical vein. To the left and right is waste rock.
- The Contractor's Trick: The operator widens the pit face by scooping 1 meter into the waste wall along with the ore.
- The Result: The bucket is filled faster, and the tonnage increases.
- Audit Term: "Edge Dilution."

Stage 3: The Stockpile (Blending Fraud)

- The Scenario: The contractor has mined a pile of "Low Grade" or Waste material that they won't get paid for (or paid less).
- The Contractor's Trick: They dump one truckload of waste in between four truckloads of good ore at the Crusher stockpile.
- The Result: The crusher mixes it all up. The waste "disappears" visually, but the final output grade drops.

Audit Test	How to Execute
1. The "Assay Penalty" Check	Review the Daily Assay Report for the specific stack mined by the contractor. Formula: If Actual Grade < Cut-off Grade (e.g., <55% Fe), check the contractor's bill. The Check: Did Finance deduct the "Dilution Penalty"? (Standard contracts penalize the *entire* batch rate if dilution > 5%).
2. The Visual "Floor" Check	Go to the pit immediately after the contractor has cleared a bench. The Check: Look at the floor. Is it smooth and hard (rock)? or is it dug up, uneven, and showing a different color (soil/clay)? Red Flag: If the floor is "gouged," they dug too deep.
3. The "Visual Band" Check	Look at the pit face (the wall they are digging). The Check: You should see clear color bands (e.g., Red for Iron Ore, White for Shale/Waste). Red Flag: If the excavator marks clearly go into the "White" zone, they are mining waste and calling it ore.

6.4 Other Red Flags & Fraud Indicators

- The High number of misfires: explosives that didn't go off, which are then claimed to be destroyed. Often, these are stolen.
- Uniform Cycle Times: If the logbook shows every trip taking exactly "45 minutes," the data is faked. Real mining has variance.
- Rising Powder Factor, Falling Production: You are using more explosives to produce less rock. This is a classic indicator of theft.
- Queuing at the Crusher: Indicates "Over-trucking." The mine is paying for too many trucks to sit idle in line.

6.5 Technical Insight: The "Boulder" Argument

This is a common dispute mechanism.

Scenario: The Crusher Manager complains that the "Raising Contractor" is sending too many oversized boulders, causing jams.

Contractor's Defense: "The blasting team (Mine Owner) didn't use enough powder."

Audit Test: Check the "Secondary Breakage" cost. If the mine is spending huge amounts on rock breakers (hydraulic hammers) after the blast, the primary blasting process is failing.

6.6 Case Study: The "Diesel Drain"

Brief Case Study: A "Raising Contractor" had a clause allowing pass-through of diesel costs. The auditor noticed the contractor’s trucks were showing fuel efficiency of 15 Litres/Hour, whereas the manufacturer standard was 45 Litres/Hour.

The Fraud: The contractor was siphoning diesel from the mine's bowser for their private fleet outside the mine, and billing the mine for "efficient" consumption.

Chapter 7: Heavy Earth Moving Machinery & Maintenance

7.1 Introduction & Audit Objective

HEMM (Dumpers, Excavators, Dozers, Graders) represents the largest capital asset on the balance sheet. Maintenance costs usually constitute 30-40% of the total mining cost.

The Internal Auditor's objective is to verify that these assets are not just "running," but running efficiently. A machine can be operational but bleeding money through excessive spare parts consumption or fuel inefficiency. For the auditor, this chapter is about Asset Governance. The risks here are not just about "Breakdowns"; they are about Capital Destruction—buying machines that sit idle, repairing machines that should be scrapped, and scrapping machines that are being stolen piece by piece.

The Value Chain: Machine Scheduling > Operations > Breakdown/Preventive Maintenance > Spares Consumption > Scrap/Disposal

Primary Audit Risks:

- Low Availability: Assets sitting idle due to poor maintenance planning (breakdown mode).
- Warranty Leakage: Paying for repairs on engines/parts that are still under warranty.
- Cannibalization: Stripping parts from a "sick" machine to fix another, turning the first machine into a skeleton (Total Loss).
- Tire Mismanagement: OTR (Off-The-Road) tires are incredibly expensive. Poor road conditions destroy them early.

7.2. Critical Field Definitions

Before entering the workshop, the auditor must understand specific terminology used in asset management and safety.

- Rotables: High-value spare parts (Engines, Transmissions, Alternators) that are not thrown away when they fail. They are sent for "Refurbishment," repaired, and returned to stock.
- Cannibalization: The practice of removing a working part from a "sick" machine to fix another machine. While sometimes necessary, it often leads to the donor machine becoming a permanent "skeleton" (Total Loss).
- PCR (Planned Component Replacement): The target life of a component (e.g., an Engine should last 12,000 hours). Replacing it too early is waste; replacing it too late risks catastrophic failure.
- Hot Work Permit: A safety license required for any maintenance involving open flames, welding, or sparks. Critical for preventing fires and protecting electronic components (ECUs).
- Cold Work Permit: A safety license for hazardous non-fire work, such as high-pressure hydraulic repairs or working at heights.
- LOTO (Lock Out Tag Out): A safety protocol where the power source (battery/starter) is physically locked with a padlock to prevent accidental startup during repair.

7.3 Key Audit Focus Areas

A. The "Availability" vs. "Utilization" Matrix

The auditor must validate the reported KPIs.

- Mechanical Availability (MA): Is the machine capable of working?

  MA% = (Scheduled Hours - Maintenance Hours)x 100 /Scheduled Hours

- Physical Utilization (PU): Is the available machine actually used?
- Audit Check:

  High Availability + Low Utilization = Over capitalization (We bought too many machines).

B. Maintenance Strategy: Breakdown vs. Preventive

- Breakdown Maintenance: Fixing it when it breaks (High cost, unplanned downtime).
- Preventive Maintenance (PM): Scheduled servicing (Oil change, filter change).
- Predictive Maintenance (PdM): Using physical inspection/sensors/oil analysis to predict failure before it happens.
- Audit Check: Review the "Backlog" of maintenance work orders. If the backlog is rising, the mine is entering a "death spiral" of breakdowns.

C. OTR Tire Management

Tires are the "black gold" of mining opex.

- TKPH (Tonne-Kilometre Per Hour): The load rating of the tire.
- Scrap Analysis: Why are tires failing?
  - *Cut Separation:* Bad haul roads (Operations fault).
  - *Heat Separation:* Speeding or Overloading (Operations fault).
  - *Tread Wear:* Normal life (Acceptable).

7.4 Detailed Audit Work Program (Checklist)

Section 1: Operational Efficiency (Utilization, Fuel, Tires)

Focus: Are the assets working efficiently or burning cash?

Sub Process	Red Flag / Risk	Audit Procedure (Field Steps)
1. Utilization Logic	The "Parking Lot" Risk: Buying expensive machines that sit idle. High Availability (90%) but Low Utilization (40%) means Capex was wasted.	The "Avail-Util" Matrix: Pull the monthly Fleet Report. Compare Available Hours vs. Worked Hours. If a dumper was "Available" for 600 hours but "Worked" only 200 hours, ask why. Is the mine "Over-trucked"?

2. Hour Meter Integrity	Phantom Utilization: Operators inflating "Hour Meter Readings" (HMR) to improve efficiency KPIs or hide long breaks.	Physical HMR Verify: For sample machines, climb into the cabin. Read the digital/analog HMR. Compare it strictly with the Logbook and FMS reading. Higher discrepancies indicate data fraud.
3. Micro-Stoppages	Hidden Inefficiency: Machines stopping for 10 minutes every hour (smoke breaks, waiting), which manual logbooks never record.	FMS "Idle" Analysis: Extract the "Wait Time" report from the Fleet Management System (FMS). Look for patterns of "Micro-stoppages" that accumulate to significant lost production.
4. Tire Governance	Pressure Neglect: Running tires under-inflated generates heat and causes "Heat Separation," destroying an expensive asset.	Cold Pressure Audit: Visit the "Tire Bay" at shift start. Gauge 3 random dumpers. Compare with OEM specs. A variance >10% significantly reduces tire life.
5. Engine Health	Impending Failure: An engine burning excessive lube oil is about to seize.	Fuel-to-Lube Ratio: Calculate Lube Oil Consumed / Fuel Consumed. If this ratio is rising month-on-month, it indicates "Blow-by" (worn piston rings). Flag for immediate inspection.

Section 2: Maintenance Quality & Spares Governance

Focus: Are we fixing it right, or just patching it up?

Sub Process	Red Flag / Risk	Audit Procedure (Field Steps)
6. Maintenance Strategy	Premature Failure: Engines failing at 8,000 hours when their "Target Life" (PCR) is 12,000 hours.	PCR Life Tracking: Track the last 5 engine replacements. Calculate the Average Life Achieved. If it is consistently 30% below OEM standards, maintenance practices are failing.
7. Oil Analysis (SOS)	Fake Overhauls: The workshop claims to have overhauled an engine, but they actually just painted it and changed the oil.	SOS Trend Review: Review "Scheduled Oil Sampling" (SOS) reports. If an engine was "Overhauled" but the very next sample shows high Iron/Wear Metals, the overhaul was fake.
8. Repeat Failures	Root Cause Ignorance: Replacing the same hose every week because the mechanic ignores the real issue.	Repeat Failure Analysis: Filter the Maintenance Log by "Machine ID" and "Part Name." If "Hose Pipe" appears 4 times in a month for the same machine, report it as root cause neglect.
9. The "Lemon" Asset	The "80/20" Trap: 20% of the fleet consumes 80% of the budget.	MTBF Outlier Analysis: Calculate Mean Time Between Failures for the fleet. Identify outliers (e.g., Dumper #109 has MTBF of 12 hours vs. fleet avg 50 hours). Recommend disposal.
10. Spares Quality	The "Counterfeit" Economy: Buying "L1" cheap spares that fail fast.	Consumption Velocity Test: Compare consumption of Cheap vs. OEM brands. Calculate "Cost per Hour". Often, the expensive

		OEM part is cheaper per operating hour.

Section 3: Refurbishment & Rotables (The "Black Hole")

Focus: Tracking high-value items sent out for repair.

Sub Process	Red Flag / Risk	Audit Procedure (Field Steps)
11. Rotables Tracking	The "Lost Asset" Risk: A damaged Engine sent to a vendor 6 months ago never came back.	Returnable Gate Pass Trace: Select sample assemblies removed > 3 months ago. Trace: Machine > Scrap Yard > Gate Pass > Vendor. Is it physically at the vendor or back in Store?
12. Repair Quality	The "Spray Paint" Overhaul: Vendor charges hefty for an overhaul but only cleans and paints it.	Test Bench Verification: Demand Post Repair Diagnosis/Dyno/Pressure Test Reports of the item. Verify serial numbers match.
13. Inventory Re-Entry	Invisible Inventory: Refurbished parts returning to the mine but sitting on the floor (not in ERP).	GRN Verification: Was a "Material GRN" created upon return? Item must appear in "Serviceable Stock" with a distinct code (e.g., REF-ENGINE).
14. Asset Identity	Identity Laundering: Vendor swapping your High Value Core with a Scrap Core.	Serial Number Match: Does the Serial Number of the item *sent out* match the Serial Number of the item *returned*? If plates are missing/changed, the asset was swapped.

Section 4: Strategic Decisions (Lifecycle, Buy vs Repair)

Focus: Is the Company making financially sound decisions?

Sub Process	Red Flag / Risk	Audit Procedure (Field Steps)
15. Economic Repair	Beyond Economic Repair (BER): Spending >60% of new price to fix an old part.	The "60% Rule": Compare Repair Estimate vs. New Cost. If Repair > 60% of New, challenge the decision. It is often cheaper to buy new.
16. Lifecycle Decisions	The "Zombie Asset": Maintenance cost > Depreciation of new machine.	Plot Maintenance Cost per Hour for 3 years. If the exponential rise exceeds the ownership cost of a new machine, the "Repair" decision is flawed.
17. LOM Alignment	Stranded Assets: Buying new fleet (10yr life) for mine closing in 3 yrs.	LOM Match Check: Compare Asset Useful Life with Remaining Mine Life. If Asset Life > Mine Life, the decision should be Rent/Lease.
18. TCO Fuel Factor	Ignoring Efficiency: Repairing an old engine that burns 10% more fuel than a new one.	TCO Calculation: Ensure Fuel Savings are included in "Buy vs. Repair" math. Fuel savings often cover the EMI of a new machine.
19. Capital Rationing	Budget-Driven Decisions: Patching up assets because "Capex Budget" is empty.	Q4 Spending Pattern: Scrutinize high-value repairs in Feb/March. Flag as deferred liability if replacement was actually needed.
20. Inhouse vs. Outsource	Hidden Overheads: Incorrectly costing In-House vs. Outsourced repairs.	Fully Loaded Rate Test: Compare Vendor Quote vs. Internal Cost (including overheads/consumables).

		Ensure Apple-to-Apple comparison.
21. Capacity Utilization	Double Payment Risk: Paying internal staff but outsourcing jobs.	Man-Hour Check: Compare Available Internal Hours vs. Booked Hours. If low utilization + high vendor spend, investigate.
22. The "Bone Yard"	Cannibalization: Stripping parts from "sick" machines; total loss of asset value.	Scrap Yard Tour: Check machines parked >6 months. Are missing parts recorded or stolen?

Section 5: Commercial Governance (Warranty & AMC)

Focus: Preventing duplicate payments and enforcing contracts.

Sub Process	Red Flag / Risk	Audit Procedure (Field Steps)
23. Warranty Management	Leakage: Paying for repairs on parts still under warranty.	Claim Success Rate: Cross-check "Breakdown Register" (<1 yr items) with "Warranty Claims". Calculate success ratio.
24. AMC Scope Audit	The "Double Dip": Paying AMC fee + separate invoices for covered items.	Invoice vs. Contract: Cross-ref "Ad-hoc Repair" invoices with AMC Scope. Recover amounts for covered services billed separately.
25. Performance Guarantees	Unclaimed Penalties: Vendor paid full despite missing Availability/Fuel targets.	LD Calculation Check: Compare Actual vs. Guaranteed Performance. Verify if Penalty/LD was deducted from the bill.
26. End-of-Warranty Check	The "Expiry" Miss: Warranty expiring without fixing latent defects.	The "11th Month" Audit: Identify assets expiring in 30 days. Ensure a "Snag List" was raised to OEM for free rectification.

27. Spare Part Warranty	Warranty within Warranty: Paying to replace a part that failed within its own warranty period.	Purchase History Trace: Check Last Purchase Date of the failed part. If < 6 months, demand Free Replacement.

Section 6: Safety & Permit Governance

Focus: Protecting the Asset and the Insurance Claim.

Sub Process	Red Flag / Risk	Audit Procedure (Field Steps)
28. Hot Work Permits	Insurance Void: Welding near fuel/tires without a permit risks fire and insurance denial.	Permit Audit: Select 5 "Hot Work Permits." Did the Safety Officer sign them *before* work started? Retroactive signing is a violation. (Check compliance to other policy covenants)
29. LOTO (Lock Out Tag Out)	The "Crush" Risk: Accidental startup while a mechanic is inside the machine.	LOTO Verification: Walk through the workshop. Look for Red Locks & Tags on isolators. If the key is just in the mechanic's pocket, it's a violation.
30. Tire Inflation Safety	Explosion Risk: Inflating giant tires outside the safety cage.	Cage Usage Audit: Are tires being inflated inside the steel safety cage? Check air hose length (must allow safe distance). (Similarly check safe/compliant operation of other equipment)

7.5. Field Insights for the Auditor

The "Fresh Paint" Trap

When auditing Checkpoint 22 (The Bone Yard), be wary of machines that look shiny. Managers often paint a "skeleton" machine (which has no engine inside) to make it look operational during a VIP or Audit visit.

- The Test: Don't just look at the outside. Climb up and ask the operator to "Key On" (start the ignition). If it doesn't crank, it's a prop.

The "Tagging System" Check

When you visit the Workshop or Store, look at the high-value spares lying on pallets.

- Green Tag: Serviceable (Ready to use).
- Red Tag: Unserviceable (Needs repair).
- Yellow Tag: Hold (Under investigation/Warranty claim).

Audit Observation: If you see untagged rusty parts mixed with clean parts on the same rack, the mine has lost control of its Rotables inventory. Mechanics will waste hours installing a broken part by mistake.

7.6 Red Flags

- "Emergency" POs: A high volume of emergency air-freight orders for spares indicates zero planning.
- The "Clean" Logbook: If a machine is 5 years old but the logbook shows zero breakdowns for a month, the operator is likely not recording minor stops.
- Repeat Failures: The same part (e.g., hydraulic hose) failing every week on the same machine. Indicates root cause is not fixed (e.g., a bent clamp rubbing the hose).

7.7 Case Study: The "Warranty Void"

Brief Case Study: A mine purchased 10 new Dumpers. The OEM warranty required using "Genuine Filters." The Purchase Manager bought "Compatible" cheap filters to save 5000 per unit. Three engines seized (5000000 each). The OEM voided the warranty because of non-genuine parts. The "saving" of 15000 cost the company 150,00000.

(This can happen in opposite manner also, where expensive OEM may be used instead of similar quality cheaper spares, or spares ordered through Original Part Manufacturer instead of OEM. For critical and expensive spares, thorough study must be conducted to identify alternate options)

8.1 Introduction & Audit Objective

Mining inventory is unique: it is voluminous, outdoor, and shapeless. You cannot count it like boxes in a retail store. It requires Volumetric Surveying.

The Internal Auditor is the "guardian of the quantity." If the reported stockpile is 1 Million Tonnes, but physically there is only 800k Tonnes, the company's profit is overstated, and a massive "Write-off" is looming.

Primary Audit Risks:

- Phantom Stock: Booking production that doesn't exist to boost KPIs.
- Density Manipulation: Using the wrong "Bulk Density" factor to convert Volume (m^3) to Mass (Tonnes).
- Floor Cleaning: Leaving a thick layer of ore on the ground (carpet) to hide shortages, or scraping up soil to inflate tonnage. Inversely, scraping the floor mixes mud/stones with the ore, increasing tonnage (Good for Production team) but destroying grade (Bad for Sales team).

8.2 Key Audit Focus Areas

A. The Survey Method

How is the volume measured?

- Total Station: Accurate but slow. Requires climbing the pile (unsafe).
- Drone / Photogrammetry: Fast, safe, and highly accurate. Generates a "Point Cloud."
- Audit Requirement: The auditor must accompany the surveyor during the "Cut-off" measurement.

B. The "Bulk Density" Factor (The Magic Number)

Volume is Science (m^3); Tonnage is Math (m^3 x Density).

- The Risk: If a stockpile is 100,000 \m^3:
    - Density 2.0 = 200,000 Tonnes.

    - Density 2.2 = 220,000 Tonnes.
- Audit Check: A 10% change in density creates 10% fake production. The auditor must insist on a "Bucket Test" or Lab Density Test for *loose* material (not in-situ density).

C. Reconciliation: The "Mass Balance"

Opening \ Stock + Production- Dispatch- Loss = Closing \ Stock

- Book Stock: What the accounts say.
- Physical Stock: What the survey says.
- Audit Tolerance: A variance of 2% is usually acceptable (Survey error). >5% is a red flag.

8.3 Detailed Audit Work Program (Checklist)

Section 1: Physical Verification (The Stock Take)

Focus: Ensuring the volume (m^3) reported matches physical reality.

Sub Process	Red Flag / Risk	Audit Procedure (Field Steps)
1. Cut-Off Procedures	Double Counting: Trucks dumping ore onto the stockpile *during* the survey, or loading from it. This leads to the same material being counted twice or missed entirely.	Freeze Time Audit: Ensure all weighbridges and dumpers stop movement during the survey window. If operations cannot stop, record the specific Truck IDs dumping during the survey and manually adjust the final volume.
2. Base Level Preparation	Volume Inflation: The surveyor assumes the "Floor Level" is lower than it actually is. Counting 1 meter of ground soil as Iron Ore adds thousands of fake tonnes.	The "Zero Level" Check: Verify the Original Ground Level (OGL) data used in the software. Ask the surveyor to shoot the "Toe" (bottom edge) of the pile. If the Toe level is higher than the OGL used in the model, the volume is overstated.

3. Digital Contour Check	Data Manipulation: The 3D model contains artificial "Spikes" (noise) caused by conveyor belts or trees being counted as rock piles.	Model Visualization: Ask to see the "Wireframe Model" or "Point Cloud" on the surveyor's screen. Look for unnatural spikes or floating points. Ensure "Non-Stockpile Objects" (conveyors, retaining walls) are digitally removed from the calculation.
4. The "Carpet" Trick	Hiding Shortages: The stockpile is spread out over a huge area (low height) rather than a high heap. A 10cm error over 10 hectares creates massive fake volume.	Height-to-Area Analysis: If a stockpile covers a large area but has a low average height (< 2 meters), flag it as high risk. Demand a Total Station survey instead of a Drone survey for such "Carpets" to ensure accuracy.
5. Dead Stock Analysis	Phantom Assets: "Dead Stock" at the bottom of the pile that has hardened into concrete or (for coal) turned into ash due to spontaneous combustion.	Aging Report: Request the "Stockpile Ageing Report." If material has sat for > 1 year, physically inspect the base. If it's fused/contaminated, insist on a Provision for Obsolescence.

Section 2: Valuation & Quality (The "Grade" Audit)

Focus: Ensuring the density and grade applied to the volume are correct.

Sub Process	Red Flag / Risk	Audit Procedure (Field Steps)
6. Bulk Density Validation	Tonnage Inflation: Using a density factor of 2.2 when the actual loose density is 2.0. A 10% density hike creates 10% fake production.	The "Bucket Test": Do not accept "Standard Factors." Witness a density test on-site. Fill a calibrated box or a dumper with the *current* loose material and weigh it. Formula: Net Weight / Known Volume.

7. Moisture Correction	Paying for Water: During monsoons, stockpiles absorb 10-15% moisture. If the survey volume is converted using "Dry Density," the inventory is massively overstated.	Free Moisture Adjustment: Check if the "Rainy Season Protocol" is active. Ensure the density factor is adjusted for Free Moisture. If the lab says 12% moisture, the inventory weight must be discounted accordingly.
8. Grade Reconciliation	Financial Misstatement: The books say the stock is "64% Fe" (High Value), but it is actually mixed with waste.	The "Grab Sample" Test: Accompanied by a geologist, take a grab sample from the stockpile. Send it to an External Lab. If the result is 60% Fe vs. Book Stock 64% Fe, the inventory value on the Balance Sheet is wrong.
9. Contamination Audit	Grade Destruction: The loader operator digging too deep and scraping the soil floor to inflate dispatch tonnage.	Silica/Alumina Trend: Check the assay reports for the bottom 10% of the stockpile dispatch. A sudden spike in Silica (SiO_2) indicates the floor soil is being mixed with the ore.
10. Mass Balance Check	Theft / Leakage: The classic "Equation of Truth" doesn't balance.	Reconciliation Formula: Calculate Opening Stock + Production - Dispatch - Loss = Closing Stock.

8.4 Red Flags & Fraud Indicators

- The "Carpet" Trick: The stockpile is spread out over a huge area rather than a high heap. This makes it harder to survey and easier to hide volume errors (a 10cm error over 10 hectares is massive).
- Perfect Matches: If Book Stock is 145,670 Tonnes and Physical Survey is 145,670 Tonnes, it is fake. Physical surveys *always* have variance.

- Changing Densities: The density factor used for the same material changes from month to month to smooth out production figures.

8.5 Case Study: The "Invisible Mountain"

Brief Case Study: A coal mine reported 1 Million Tonnes of stock. The auditor noticed the stockpile area was lush green with grass growing on it (indicating no movement). The Reality> A drone survey revealed the "hills" were actually overburden covered with a thin layer of coal (veneer). The actual coal stock was zero. The manager had been reporting fake production to get bonuses.

Part III: Value Addition & Utilities (The Plant)

Chapter 9: Ore Beneficiation & Refining (The Black Box)

The Value Chain: Crushing > Grinding (Milling) > Separation (Floatation/Gravity) > Thickening (Dewatering) > Filtration > Concentrate Stockpile

9.1 Introduction & Technical Context

The Beneficiation Plant (Concentrator or Washery) is often treated as a "Black Box" by financial auditors. Ore goes in, Concentrate comes out, and a massive amount of slurry (Tailings) is discarded.

The Auditor's Challenge:

Unlike a warehouse where 10 items in = 10 items out, a plant involves chemical separation. The auditor must understand Metallurgical Accounting. If the plant reports a recovery of 85% when the theoretical maximum is 92%, that 7% variance represents millions of dollars in lost metal or theft.

9.2 Critical Technical Concepts & Formulas

A. The Two-Product Formula (The Auditor's Weapon)

You do not need to weigh the tailings (which is difficult) to know if the production report is accurate. You only need the Assay Grades from the lab.

Recovery (R)% = (c(f-t))/(f(c-t)) X 100

- f = Feed Grade (% Fe/Au/Cu entering the plant)
- c = Concentrate Grade (The final product)
- t = Tailings Grade (The waste)

A practical illustration of how an auditor uses assay grades to catch fraud without needing to weigh the waste.

The Claim: The Monthly Production Report states the plant achieved a 94% Recovery Rate. (Meaning: 94% of the iron entering the plant ended up in the final product).

The Audit Data (The 3 Lab Numbers)- Check assay report from the lab:

Feed Grade (f): 62.0% Fe (The quality of rock entering the crusher).

Concentrate Grade (c): 64.5% Fe (The final product quality).

Tailings Grade (t): 22.0% Fe (The waste).

The Calculation (The "Two-Product" Formula)

R% = (64.5(62-22)) / (62(64.5-22)) x100=97.9%

The Audit Conclusion

Calculated Recovery: 97.9%

Reported Recovery: 94.0%

The Variance: ~4%

Interpretation: The plant is more efficient than they are reporting. A 4% gap in a large mine is massive.

- Possibility A (Theft): The plant actually produced more concentrate than recorded, and the extra 4% (the "gap") was siphoned off or sold without invoicing.
- Possibility B (Data Manipulation): The "Tailings Grade" (t) of 22% might be wrong. If the real tailings grade was higher, the recovery would drop. The plant might be intentionally under-reporting the tailings grade to hide inefficiency.

Auditor's Action: You have caught a "Metallurgical Imbalance." You must now investigate Possibility A (Check dispatch records for unauthorized trucks) or Possibility B (Send a tailings sample to an external lab to verify if it is truly 22%).

B. Grinding Media Consumption (The "Steel Ball" Audit)

Mills (Ball Mills/SAG Mills) use steel balls to crush rock. These balls wear down and are a massive consumable cost.

- Metric: kg of steel balls consumed per Tonne of ore milled.
- The Fraud: Operators selling new steel balls to scrap dealers and claiming they were "consumed" in the mill.
- The Test: Perform a "Ball Drop" test or analyze the "Ball Kibble" addition logs vs. mill operating hours. (can refer an independent periodic “ball drop” test).

C. Mine-to-Mill Reconciliation (The "F-Factors")

The auditor must check the Call Factors that reconcile the three versions of "Truth":

1. F1 Factor (Reserve vs. Grade Control): Did the block model predict the ore accurately before digging?
2. F2 Factor (Mine vs. Plant): Did the grade sent by the Mine match the grade received by the Plant?
3. F3 Factor (Product vs. Sales): Did the final production match what was sold?

- Audit Step: Calculate variance for F2.
- Red Flag: If F2 is consistently < 0.98 (i.e., Plant receives less metal than Mine claims to send), the Mine Operations team is likely mining waste and blending it with ore to hit volume targets ("Dilution"), while claiming full ore grade.

Unlike a car factory where you can count half-assembled cars, a mining plant has "Gold/Metal in Circuit" (MIC)—dissolved gold or concentrate slurry stuck in tanks and pipes. This "WIP" can be worth millions. Managers often manipulate this "estimate" to smooth out monthly production figures.

9.3 Detailed Audit Work Program: Plant Operations & Quality Control

Section 1: Metallurgical Balance & Yield (The "Black Box" Audit)

Focus: Verifying if the metal entering the plant actually comes out as product.

Sub Process	Red Flag / Risk	Audit Procedure (Field Steps)
1. The Two-Product Check	The "Invisible Loss": The plant claims 90% recovery, but actual recovery is 85%. The difference is stolen	The Formula Test: Do not trust the Production Report. (Explained in detail above)

	concentrate or unreported waste.	
2. Tailings Dam Audits	The "Golden Tailings": Intentionally suppressing the reported Tailings Grade (e.g., reporting 8% Fe when it is 14% Fe) to hide process inefficiency or theft.	External Sampling: Accompanied by an external chemist, go to the final discharge pipe. Take 3 grab samples. Send them to an external lab (e.g., SGS/Bureau Veritas). If external lab result > internal lab result, the plant is manipulating data.
3. Recirculating Load	The "Ghost Volume": Increasing the "Recirculating Load" data in the SCADA system to hide missing mass (stolen concentrate) by pretending it is still "looping" inside the plant.	Power vs. Load Check: Check the Mill Power Consumption (Amps). If the SCADA shows a massive load circulating, but the Mill Power is low/normal, the load is fake.
4. Mine-to-Mill (F2)	The "Dilution" Gap: The Mine claims to send 60% Fe, but the Plant receives 58% Fe. The "F2 Factor" (Plant Receipt / Mine Dispatch) drops below 0.98.	The "F2" Trend Analysis: Plot the F2 Factor for the last 6 months. If it is consistently < 1.0, the Mine Operations team is mining waste and blending it with ore ("Dilution") to hit tonnage targets, while the Plant takes the blame for "Yield Loss".
5. Metal in Circuit (MIC)	Income Smoothing: Managers manipulating the "Work in Progress".	Dip & Density Check: Challenge the month-end WIP. Ask for the "Tank Level Dips" and "Slurry Density" logs at 11:59 PM on the

	"Draining tanks" to boost month-end production or "filling tanks" to hide excess.	cutoff date. A high, flat MIC figure for 3 months is a red flag.

Section 2: Plant Inputs & Efficiency

Focus: Ensuring the inputs (Feed, Reagents) are genuine.

Sub Process	Red Flag / Risk	Audit Procedure (Field Steps)
6. The "Salting" Check	Feed Grade Inflation: Operators manually tossing high-grade lumps into the sample bucket at the feed belt to create a false high Feed Grade (masking plant inefficiency).	Visual Surveillance: Stand at the Auto-Sampler location for 1 hour. Ensure no manual intervention occurs. Check if the "Sample Cutter" cuts the *entire* stream or just the top layer (which might be biased).
7. Moisture Arbitration	The "Water Tax": Ore is sold on a "Dry Metric Tonne" (DMT) basis. Overstating moisture by 1% reduces the billable weight (Revenue) by 1%.	Oven Verification: Witness the moisture determination process. Ensure the sample is weighed *immediately* after collection and *immediately* after coming out of the oven. Any delay causes evaporation and under-reporting of moisture.
8. Weighometer Integrity	Production Inflation: The Belt weigher reads 100 TPH even when the belt is empty or lightly loaded due to calibration drift.	The "Chain Test": Request a "Zero Calibration" (run belt empty) and a "Chain Test" (run a known weight chain). If the error is > 0.5%, all tonnage reported since the last calibration is suspect.

Section 3: Consumables & External Processing

Focus: Cost control and preventing theft of supplies.

Sub Process	Red Flag / Risk	Audit Procedure (Field Steps)
9. Grinding Media (Balls)	The "Scrap" Scam: Operators selling new steel balls to scrap dealers and claiming they were "consumed" in the mill to explain the shortage.	The "Ball Drop" Test: Calculate the Specific Consumption (kg/Tonne). If it spikes, conduct a "Marked Ball Test" (drop tagged balls and measure wear) or reconcile the "Ball Kibble" addition logs vs. Mill Operating Hours.
10. Reagent Dosage	Chemical Theft: Expensive chemicals (Frothers/Collectors) being stolen. Pumps are set to low dosing, but inventory records show high consumption.	Pump vs. Tank Check: Compare the "Pump Stroke" setting (SCADA) with the actual "Level Drop" in the Day Tank. (Check actual consumption with SCADA readings.)
11. Liner Wear Profile	Kickbacks: Changing Mill Liners (expensive steel plates) prematurely (e.g., at 50% wear) to generate more orders for a specific vendor.	Scrap Inspection: Visit the scrap yard. Inspect the discarded Mill Liners. Measure the thickness. If they are not worn down to the shell/critical thickness, investigate the Change-out Decision.
12. Job Work (Tolling)	Value Leakage: Third-party washeries keeping the high-grade "Slimes" or "Rejects" which still contain recoverable metal.	Mass Balance of Job Worker: Reconcile Sent Tonnage = Product Received + Reject Received. Demand a "Disposal Certificate" for the rejects. Test the "Reject" grade—if it's high value, the washery is stealing it.

9.4 Fraud Case Study: The "Recirculating Load" Trick

A Gold processing plant was stealing gold concentrate. To hide the missing mass, they increased the "Recirculating Load" (material sent back to the start) in the SCADA system data. This made it look like the material was still "in the circuit" rather than stolen. The auditor caught this by noticing the mill power consumption didn't match the massive load theoretically circulating.

9.5 Progressive Tailings & Rehabilitation Planning

- Rehabilitation Scheduling vs. Production
  - Audit Check: Compare annual tailings deposition plan with progressive rehabilitation schedule (hectares reclaimed per year) and corresponding budget.
  - Risk: Pushing all rehabilitation to "end of LoM", leaving an unfunded closure liability.
- Tailings Re-mining & Re-processing
  - Audit Check: Where historic tailings are re-mined or re-processed, ensure metallurgical accounting covers these streams separately and royalty/tax treatment is clear.
  - Fraud Indicator: High-grade "tailings" shipped out as product without appropriate royalty or disclosure.

Chapter 10: Captive Power Plant (CPP) Operations

10.1 Introduction & Technical Context

Thermal Captive Power Plants (CPPs) are common in large-scale mining and mineral processing operations in India, especially in energy-intensive sectors like zinc, aluminum, and steel.

- Mining operations—especially those involving crushing, grinding, smelting, and refining—require continuous, high-load power.
- Grid power can be unreliable or expensive in remote mining regions, making thermal CPPs (coal, lignite, or gas-based) a practical solution.

A CPP is a thermal balance sheet. Fuel (Coal) enters, Heat is released, Steam is generated, and Turbines spin. The efficiency is measured in Heat Rate.

Value Chain: Coal Handling > Boiler (Combustion) > Turbine (Generation) > Condenser (Cooling) > Switchyard (Evacuation) > Ash Disposal

The Golden Rule:

- Design Heat Rate: ~2,250 kCal/kWh (Manufacturer promised).
- Operating Heat Rate: ~2,350 kCal/kWh (Acceptable degradation).
- Audit Red Flag: >2,600 kCal/kWh (Fuel theft or severe equipment failure).

10.2 Critical Technical Concepts

A. GCV Accounting: The "Moisture Loss" Gap

Coal value depends on Gross Calorific Value (GCV).

1. GCV (ARB): *As Received Basis* (Includes total moisture). This is what you pay the supplier for.
2. GCV (AFB): *As Fired Basis* (What goes into the boiler).

The Audit Gap: If coal sits in the yard for 3 months, it loses GCV (weathering). However, a drop of >200 kCal between "Receipt" and "Firing" usually means Good Coal was swapped for Bad Coal (Shale).

Global Case Study: The "GCV Arbitrage"

- Scenario: A Power Plant in SE Asia paid for 5,500 kCal coal based on "Load Port" reports.
- Fraud: Suppliers performed a "Mid-Sea Swap," replacing 20% of the cargo with dirt.
- Audit Find: Using a "Reverse Heat Balance", the auditor proved the actual coal burned was only 4,800 kCal. The company was paying for premium coal but burning dirt.

B. Specific Oil Consumption (SOC)

Boilers need Light Diesel Oil (LDO) only for startup. Once the fire is stable, oil is cut off.

- Benchmark: < 0.5 ml/kWh.
- The Issue: If SOC is > 2.0 ml/kWh, operators are running oil support constantly because they are feeding bad quality coal (wet/stones) that kills the flame. This is a double loss (Expensive Oil + Bad Coal).

C. The PAT Scheme (Regulatory Energy Audit)

In India, CPPs are "Designated Consumers" under the Bureau of Energy Efficiency (BEE). They must meet specific "Net Heat Rate" targets.

- The Asset: ESCerts (Energy Saving Certificates). If the plant is efficient, it earns ESCerts which are tradeable assets (Revenue).
- The Liability: If the plant misses the target, it must *buy* ESCerts to comply.
- Audit Check:
    1. Verify the "Form 1" submission to BEE. Does the reported energy consumption match the SAP Consumption?

2. Check if the value of ESCerts held is recognized in the books (Ind AS 20- Govt Grants) or if a provision is made for the shortfall liability.

10.3 Detailed Audit Work Program: Captive Power Plant (CPP)

Section 1: Fuel Management (Coal & Oil)

Focus: The largest cost component. Protecting the Calorific Value (GCV).

Sub Process	Red Flag / Risk	Audit Procedure (Field Steps)
1. Quality Verification	The "Mid-Sea Swap": Supplier delivering dirt/shale instead of coal. Using a "Friendly Lab" to pass the sample.	The "Third Sample" Test: Standard contracts have 3 referee samples. Seize the "Referee Sample" (stored in joint custody) and send it to a neutral, external lab. If the result deviates > 100 kCal from the payment grade, investigate collusion.
2. Transit Loss	The "Water Weight" Fraud: Trucks adding water before the weighbridge to make up for stolen coal. Paying for water at the price of coal.	Transit Loss Calculation: Compare Dispatch Weight vs. Receipt Weight. Standard tolerance is 0.8%. If loss is > 2%, check moisture levels. High moisture + High weight = Water added to hide theft.
3. Rejection Analysis	Paying for Stones: High volume of stones/shale rejected by the Coal Mills.	Mill Rejection Analysis: Check the quantity of "Rejects" cleared from the mill. If rejection > 1% of feed, calculate the financial loss. Trigger the "Stone Penalty" clause in the coal supply contract.

4. Blending Logic	Unscientific Blending: Mixing high-grade and low-grade coal without checking Ash Fusion Temperature (AFT), causing boiler "clinkering" (choking).	AFT Consistency Check: Review the daily lab reports for the blended mix. If AFT varies significantly between batches, the blending is unplanned/random, risking boiler tube failure.
5. GCV Loss (Yard)	Theft Disguised as Weathering: Coal sitting in the yard loses energy (weathering), but a drop > 200 kCal usually means good coal was swapped for bad coal inside the plant.	"As Received" vs. "As Fired": Compare the GCV (ARB) from the weighbridge sample vs. the GCV (AFB) from the boiler feed belt. If the gap > 150-200 kCal, investigate the "Coal Yard Management"[5].

Section 2: Plant Efficiency & Operations

Focus: Converting Heat to Power with minimal loss.

Sub Process	Red Flag / Risk	Audit Procedure (Field Steps)
6. Heat Rate Performance	The "Waiver" Fraud: The O&M contractor burns more coal than the "Guaranteed Heat Rate" but claims "Poor Coal Quality" to avoid the penalty.	Penalty Re-calculation: Re-calculate the Heat Rate Penalty monthly. Verify if the coal quality *actually* deviated from the contract specs. If coal was within specs, reject the waiver and deduct the penalty.
7. Specific Oil Consumption (SOC)	Flame Support Waste: Operators running Oil Support (LDO) constantly because they are feeding wet/bad coal, burning	Check if SOC is greater than standard. If yes, check the "Coal Mill Outlet Temp." If low, they are using oil to compensate for wet coal. Calculate the cost impact.

	expensive oil unnecessarily.	
8. Condenser Vacuum	Efficiency Killer: Low vacuum (back pressure) reduces turbine efficiency, directly increasing coal burn.	DCS Log Review: Review the trend of "Condenser Vacuum." If it is consistently poor, check the "Cooling Tower" performance. Dirty cooling water is often the root cause.
9. Auxiliary Power	Energy Waste: Fans and Pumps running at 100% speed with dampers closed instead of using VFDs (Variable Frequency Drives).	Aux Power Audit: Check the energy meters of BFP (Boiler Feed Pump) and ID/FD Fans. If Aux Consumption > 9-10%, the plant is bleeding power internally.
10. PAT Scheme Compliance	Statutory Liability: Missing the "Net Heat Rate" target set by the Bureau of Energy Efficiency (BEE), leading to a liability to buy ESCerts.	Form 1 Verification: Verify the data submitted to BEE. Check if the Shortfall Liability (Cost of ESCerts needed) is provided for in the books (Ind AS 37).

Section 3: Maintenance & Commercials

Focus: Spares, Waste Disposal, and O&M Contracts.

Sub Process	Red Flag / Risk	Audit Procedure (Field Steps)
11. Fly Ash Commercials	"Double Dipping": Paying a contractor to remove ash, while the contractor secretly sells that same ash to a cement plant.	Disposal Contract Review: Verify the "Ash Disposal" contracts. Check the local market rate. If Cement plants are buying ash, the mine should *sell* it or give it free, not *pay* for removal.

12. Refurbishment Risks	Catastrophic Risk: Refurbishing critical rotating parts (Turbine Blades) at local workshops to save cost, risking total turbine failure.	Certification Check: Verify if the "Life Extension Certificate" for refurbished rotable parts comes from the OEM (Original Equipment Manufacturer). Reject certificates from local generic workshops.
13. Rotable Tracking	The "Lost Asset": Expensive rotors or impellers sent for repair years ago and never returned.	WIP Aging Analysis: Track items moved from Store > Plant > Vendor. If an asset is at the vendor for > 12 months, it is likely lost.

10.4 Field Insight: The "Reverse Heat Balance"

If you suspect the "Coal Quality" reports are fake (i.e., the lab is lying about GCV to pass bad coal), use the Reverse Heat Balance method[16]:

1. Take the Total Power Generated (kWh).
2. Multiply by the Design Heat Rate (e.g., 2,300 kCal/kWh).
3. Result: Total Heat Consumed (kCal).
4. Divide by Total Coal Tonnage fed to boiler.
5. Result: Actual GCV Fired.

Audit Test: If the Lab says GCV is 4,000 kCal, but your reverse calculation shows 3,200 kCal, the lab is manipulating samples.

10.5 **Expert Insight:** The "Refurbishment" Trap

- The Scenario: A Critical Spares vendor offers to "Refurbish" a cracked Gas Turbine blade for 40% of the new price.

- The Risk: Without advanced metallurgy tests (crystallography), micro-cracks remain. The refurbished blade fails at 3,000 RPM, destroying the entire turbine.
- Audit Check: Verify if the "Life Extension Certificate" for refurbished parts comes from the OEM or just a local workshop. Never accept critical rotating critical spares refurbished by non-certified local vendors.

10.6 Global Case Study: The "Reverse Engineered" Turbine Blade

- Scenario: A plant in Eastern Europe bought "compatible" turbine blades from a local vendor to save $2.5M vs the OEM.
- Result: The local vendor didn't understand the harmonic frequency. A blade sheared off at 3,000 RPM, destroying the turbine.
- Lesson: "Cost Saving" on critical high-speed rotables is a catastrophic risk.

11.1 Introduction & Technical Context

The Value Chain: Water Intake > Process Usage > Effluent Collection > Treatment (ETP/STP) > Recycling (ZLD) > Sludge Disposal

Modern mines operate under Zero Liquid Discharge (ZLD) permits. This means not a single drop of process water can leave the gate.

The Cost Reality: ZLD is expensive. It requires running Reverse Osmosis (RO) and Multi-Effect Evaporators (MEE).

- Cost to Treat: INR 180 - 350 per m^3.
- Cost to Dump (Illegal): 0.
- The Incentive: Managers bypass the ETP to save power and chemicals, risking plant closure.

11.2 Critical Technical Concepts

A. The MEE (Multi-Effect Evaporator) Trap

The MEE boils reject water to separate salt. It is a massive steam guzzler.

- Steam Economy Ratio: kg of Water Evaporated / kg of Steam Consumed. (Standard = 4:1 or 5:1).
- Audit Test: Check the Steam Flow Meter to the MEE.
    - *Scenario:* Production is 100%, but MEE Steam flow is near zero.
    - *Conclusion:* The high-TDS toxic water is being bypassed (pumped into ground aquifers or storm drains).

B. Chemical Dosing: The "Jar Test"

ETPs need Lime, Alum, and Polyelectrolytes to settle sludge.

- The Audit: Ask the chemist to perform a "Jar Test" in front of you to determine the optimum PPM (Parts Per Million) of chemical required to settle the sludge.

- The Calculation: Expected \ Consumption = Flow (m^3) X Jar Test PPM

Compare this with the Store Issue Record. If Store Issue is 2x the Expected Consumption, chemicals are being stolen.

C. Tailings Storage Facility (TSF) Governance

The TSF holds millions of tonnes of toxic sludge behind an earthen dam. Failure results in loss of life and total mine closure.

- The Risk: Raising the height of the dam wall ("Embankment Raise") without proper compaction or safety permits to save costs.
- Audit Check:
    1. Piezometer Readings: Check the "Pore Pressure" logs of the dam walls. Are they being monitored daily? High pressure = Liquefaction risk.
    2. Freeboard Verification: Ensure the distance between the slurry level and the top of the dam wall matches the "Design Freeboard" (usually 1.5m). Using the freeboard space for extra storage is a major violation.

11.3 Detailed Audit Work Program: ETP & Water Management

Section 1: Zero Liquid Discharge (ZLD) & Water Balance

Focus: Detecting illegal discharge of untreated water.

Sub Process	Red Flag / Risk	Audit Procedure (Field Steps)
1. The MEE Steam Check	The "Cold Boiler" Fraud: The Multi-Effect Evaporator (MEE) is the most expensive part of ZLD (boiling water). If Production is 100% but MEE steam consumption is near	Steam Economy Ratio: Calculate kg of Water Evaporated / kg of Steam Consumed. The standard is 4:1 or 5:1[1]. If the steam flow meter shows negligible usage while the plant runs full load, the

	zero, toxic water is being bypassed.	effluent is being dumped, not evaporated.
2. Mag-Meter Integrity	Falsified Flow Data: Operators tampering with the "Magnetic Flow Meters" to under-report the volume of effluent generated or recycled.	"Empty Pipe" & Totalizer Audit: Check the "Empty Pipe Detection" alarm logs. Operators often cut the signal wire or manipulate the grounding ring to make the meter read "Zero Flow" while liquid is actually flowing. Verify the Totalizer reading against the daily logbook.
3. The "Rainy Day" Discharge	Illegal Dumping: Using heavy rain to mask the discharge of toxic effluents into storm water drains.	Night Surprise Check: Visit the final discharge point/storm drains at 2 AM or during heavy rainfall. Illegal discharge almost always happens at night or during rain to dilute the evidence.
4. RO Membrane Autopsy	Poor Pre-Treatment: Replacing expensive RO membranes every 6 months instead of 2 years. This indicates the Sand Filter/Carbon Filter is failing or being bypassed.	Replacement History Review: Review the Store Issue Register for RO Membranes. Frequent changes indicate the plant is forcing dirty water through the RO to meet targets, destroying the membranes.
5. Water Positivity Check	Fake ZLD: Claiming to be "Zero Liquid Discharge" but having no record of recycled water returning to the process.	Recycle Loop Trace: Trace the "Permeate" line from the RO plant. Does it physically go back to the Process Water Tank? Check the flow meter on the

		return line. If it reads zero, the "treated" water is being dumped.

Section 2: Chemical Consumption & Process Efficiency

Focus: Cost control and preventing chemical theft.

Sub Process	Red Flag / Risk	Audit Procedure (Field Steps)
6. The "Jar Test" Audit	Chemical Theft: The store issues 1,000kg of Lime/Alum, but the theoretical requirement is only 500kg. The rest is stolen.	The "Jar Test" Reconciliation: Ask the chemist to perform a "Jar Test" on-site to determine the optimum PPM. Calculate Expected Consumption = Flow (m3) x Jar Test PPM. Compare this with actual Store Issues. Variance > 10% suggests theft.
7. Sludge Generation Ratio	Sludge Washout: The ratio of kg Sludge / m^3 Effluent drops significantly. This means solid waste is escaping into the drains instead of being trapped in the filter press.	Mass Balance Check: Calculate the "Dry Solids" generated by the Filter Press. If the plant treated 1,000 m^3 of water but generated only 10kg of sludge (when design says 100kg), the hazardous sludge was flushed out.

Section 3: Consent To Operate (CTO) Compliance

Focus: Adherence to State Pollution Control Board (PCB) limits.

Sub Process	Red Flag / Risk	Audit Procedure (Field Steps)
8. Water Cess Reconciliation	Statutory Tax Evasion: Under-reporting water consumption to the	Meter vs. Return Check: Compare the "Actual Meter Reading" in the Daily Logbook vs. the reading

	Govt to pay lower "Water Cess" taxes.	submitted in the Water Cess Return. Any discrepancy is statutory fraud.
9. Discharge Parameters	Pollution Limit Breach: Exceeding the limits for pH, COD (Chemical Oxygen Demand), or TSS (Total Suspended Solids) set in the CTO.	Lab Report Validation: Compare the Daily Internal Lab Report against the CTO Limit Sheet. If the CTO limit for COD is 250 mg/L but the internal report shows 400 mg/L, has the plant reported this breach?.
10. Quantity Limits (Capacity)	Illegal Expansion: Treating more effluent than the licensed capacity. If the CTO permits 100 KLD (Kilo Liters/Day), processing 150 KLD is a violation.	Flow Totalizer Check: Sum the daily inlet flow to the ETP for the last month. If the daily average > Consented Quantity in the CTO, the mine is operating illegally.
11. Hazardous Waste Disposal	Illegal Dumping: Disposing of Hazardous Sludge (Category 35.3) in open pits instead of authorized TSDFs (Treatment Storage Disposal Facilities).	Manifest Audit: Verify the Form 10 (Manifest). Does the quantity of sludge "Sent Out" match the quantity "Received" by the TSDF? Check if the transporter is authorized.
12. Fly Ash Utilization	Fake Certificates: Generating fake "Utilization Certificates" from brick kilns to satisfy the MoEF requirement	Vendor Verification: Verify the brick/cement manufacturers receiving the ash. Are they real operational entities? Shell companies are often used to

	of 100% Fly Ash utilization.	generate fake compliance certificates.

Section 4: TSF (Tailings Dam) Safety

Focus: Preventing dam failure.

Sub Process	Red Flag / Risk	Audit Procedure (Field Steps)
13. Embankment Stability	Liquefaction Risk: Raising the dam wall height without monitoring the internal pressure.	Piezometer Check: Check the "Pore Pressure" logs from the piezometers installed in the dam wall. If pressure is rising, the dam is unstable. Ensure readings are taken daily.
14. Freeboard Compliance	Overflow Risk: Filling the dam to the brim to delay building a new one.	Visual Measurement: Measure the Freeboard (distance between slurry level and top of wall). It must match the "Design Freeboard" (usually 1.5m).

11.4 Case Study: The "Rainy Day" Discharge

A plant claimed ZLD compliance. However, the auditor noticed that ETP power consumption dropped to near zero during the monsoon months (July-Sept). Investigation revealed that during heavy rains, the plant opened a bypass valve, mixing toxic cyanide tailings with rainwater runoff, assuming the dilution would hide the chemicals. They saved in power but faced a fine when river fish died downstream.

Part IV: Quality & Assurance

Chapter 12: Weighbridge Operations & Anti-Fraud Controls

The Value Chain: Truck Entry > RFID Identification > Tare Weighing > Loading/Unloading > Gross Weighing > Net Weight Calculation > Challan/TP Generation

12.1 Introduction & Audit Objective

The weighbridge is the "Cash Register" of the mine. Every tonne of ore sold and every litre of fuel bought passes over these load cells.

- Inbound Risk: Suppliers inflating weights of coal/consumables (Paying for air).
- Outbound Risk: Under-weighing sales trucks (Theft/Collusion with buyers).

The Auditor's Motto: "God created the world, but the Dutch made the scales—and the thief figured out how to trick them."

12.2 The Mechanics of Weighbridge Fraud

A. Tare Weight Manipulation (The #1 Fraud)

Net Weight = Gross Weight (Loaded)- Tare Weight (Empty).

To steal ore, the trucker must increase the Tare Weight during the empty weighing or decrease the Gross Weight during the loaded weighing.

- The "Water Tank" Trick: A truck enters empty but with a hidden water tank full (or fuel tank full). They weigh the Tare (Heavy). They go to the mine, dump the water/fuel, load the ore, and weigh out.
    - *Result:* The recorded Tare is higher than the actual empty truck. The calculated Net Weight is lower. The mine gives away free ore.
    - The Audit Catch: Look for "Wet Spots" on the ground near the weighbridge or parking area where drivers dump the water. Install cameras specifically looking under the truck chassis.

- The "Double Weighing" (stored Tare): Using a "Stored Tare" from 6 months ago when the truck had new tires and a full tank, versus the current state.

B. Positioning & Sensor Fraud

- Axle Skipping: The driver stops the truck such that the rear axle is *just* off the weighbridge plate. The sensor records 90% of the weight.
- Load Cell Tampering: Installing a remote-controlled device ("The Chip") inside the junction box that reduces the digital signal by 5-10% when a button is pressed.

12.3 Detailed Audit Work Program: Weighbridge Operations

Section 1: Physical Calibration & Hardware Integrity

Focus: Is the machine weighing correctly? In addition to checking basic periodic calibration certificates perform following tests to check potential frauds.

Sub Process	Red Flag / Risk	Audit Procedure (Field Steps)
1. The "Dead Weight" Test	Linearity Error: The scale reads "0" accurately when empty but reads 19T when a 20T load is applied. This "Drift" is often intentional to steal 5-10% of the material.	Incremental Load Test: Do not just check Zero. Drive a truck (~15T) onto the scale. Record the weight. Add 1 Tonne of Certified Test Weights. The scale *must* increase by exactly 1,000 kg. If it increases by only 900kg, the calibration is manipulated.
2. Corner (Eccentricity) Test	Sensor Tampering: One specific load cell (corner) is faulty or rigged. Drivers know this and park their heavy axle on that specific corner to	The "Four Corner" Drive: Park a loaded truck on the four corners of the bridge platform sequentially. Record the weight at each spot. If the readings vary by > 20kg, the bridge is

	reduce the weight reading.	structurally compromised or the load cell is rigged.
3. Zero Error Check	Negative Zero Fraud: The operator sets the scale to "-200 kg" before the truck enters. When a 20,000 kg truck weighs in, it reads 19,800 kg.	Random "Zero" Audit: Watch the display *immediately after* a truck leaves. It must return to 0 kg (+/- 10kg). If it rests at -100kg or +100kg, the operator is manually manipulating the "Tare" or "Zero" button.
4. Sensor & Boom Logic	Axle Skipping: The driver positions the truck so the rear axle is *just off* the plate to lower the weight.	IR Beam Test: Block the optical sensor (IR Beam) at the edge of the bridge with your hand. Check if the software allows you to "Print Ticket." It should be Blocked. If the system prints while the sensor is blocked, the anti-fraud logic is disabled.

Section 2: Operational Process & Anti-Fraud

Focus: Are operators and drivers colluding?

Sub Process	Red Flag / Risk	Audit Procedure (Field Steps)
5. Tare Weight Analysis	The "Variable Tare" Fraud: A truck enters with a hidden water tank (Heavy Tare) to steal ore. Or uses a "Stored Tare" from 6 months ago.	Tare Trend: Extract data for a specific Truck ID over 3 months. If the tare fluctuates by > 300kg (fuel variance), investigate. If it fluctuates by 2 tonnes, it is the "Water Tank" trick.
6. Weighing Time Analysis	Manual Bypass: A truck is weighed in 3 seconds (physically impossible). This indicates the	Stabilization Timer Check: Filter the logs for "Weighing Duration" < 10 seconds. Real weighing takes 30-45 seconds

	operator manually typed the weight instead of letting the scale stabilize.	for the scale to stabilize. Fast times indicate data entry fraud.
7. RFID & Cloning	The "Ghost Truck": Using a cloned RFID tag. Truck A leaves the mine, but Truck B's tag is scanned.	Sequence Check: Review the "Tag Read" logs. If the *In-Weighbridge* and *Out-Weighbridge* read the same Tag ID within 1 minute (impossible transit time), two trucks are using cloned tags.
8. Camera Synchronization	Ghost Weighing: Generating a weighment slip without a truck actually being on the bridge (to create fake stock or sales).	Photo-to-Ticket Match: Select 10 random weighment slips. Retrieve the CCTV Snapshot stored in the system for that Ticket ID. *Check 1:* Is there a truck in the photo? *Check 2:* Does the Truck Number on the plate match the Ticket?

Section 3: Data Forensics & Software Controls

Focus: The digital trail.

Sub Process	Red Flag / Risk	Audit Procedure (Field Steps)
9. The "M-Flag" Audit	The Manual Override: Operators using "Maintenance Mode" or "Manual Edit" to change weight figures during the night shift.	Mode Analysis: Filter the database for records flagged "M" (Manual) or "E" (Edit). Every single manual entry must have a signed authorization slip from the Shift In-charge. A high % of M-flags (>1%) is a major red flag.

10. Integration Integrity	The "Bridge-to-Book" Gap: The Weighbridge software shows 20.5 Tonnes, but SAP/ERP shows 20.0 Tonnes. The difference is pocketed.	Database Reconciliation: Extract the raw CSV dump from the Weighbridge PC. Compare it line-by-line with the SAP Sales Register. Look for "Deleted Tickets" in the local PC that are missing from SAP.
11. Master Data (Truck Capacity)	Overloading: A 10-wheeler truck (Capacity 25T) is registered in the Master Data as a 14-wheeler (Capacity 35T) to allow overloading without triggering alarms.	RTO Verification: Pick the "Top 10 Transporters" by volume. Check their Truck Registration Certificate (RC Book). Does the Registered Laden Weight (RLW) match the capacity entered in the Weighbridge Master?
12. Software Access Control	The "Admin" Login: All 3 shifts using the same "Admin" or "Operator" login password. No accountability for fraud.	Password Post-It Test: Visit the cabin. Is the password written on a sticky note? Check the User Logs—do individual operators have unique IDs (e.g., User_John, User_Amit)?

12.4 Case Study: The "RFID Swap"

A mine used unmanned weighbridges with RFID tags on windshields. A transporter peeled the RFID tag off a 10-tonne truck and stuck it on a 20-tonne truck. The system registered the 20-tonne truck as the smaller one, allowing them to smuggle out 10 tonnes of excess material per trip without the system flagging an overload.

Chapter 13: Quality Control & Assay Lab (The Grade Audit)

This chapter addresses the "Grade Risk." In mining, the "Quantity" (Tonnage) is visible, but the "Quality" (Grade) is invisible until the lab report comes out. This makes the laboratory the "Black Box" of valuation.

The Value Chain: Sample Collection > Preparation (Crushing/Pulverizing) > Fire Assay / Titration > LIMS Entry > Certificate Generation > Sample Archival

13.1 Introduction & Audit Objective

If the Weighbridge is the *Cash Register*, the Assay Lab is the *Price Tag*.

- Iron Ore: 62% Fe sells for 6000. 58% Fe sells for 4000. (illustrative)
- Gold: 3g/tonne is profit. 1g/tonne is a loss.

The objective is to ensure the "Chain of Custody" is unbroken and the lab equipment is not "drifting."

13.2 Key Technical Concepts

A. Sampling Theory

The biggest error is not in the *Analysis* (Chemistry), but in the *Sampling* (Collection).

- The Cone & Quartering Method: The manual way of splitting a sample to ensure it is representative.
- Audit Check: Watch the sampler. Do they take the "fines" (dust) and the "lumps" (rocks) equally? Gold often hides in the fines; Iron often concentrates in the lumps.

B. QA/QC Protocols

Every batch of samples sent to the lab must contain:

1. Standard: A sample of known grade (e.g., a Certified Reference Material). *Tests Accuracy.*
2. Blank: A sample of waste rock (0% metal). *Tests Contamination.*

3. Duplicate: The same sample split into two. *Tests Precision.*

C. Critical Commercial Risk: "Moisture Determination" (Moisture Trap)

- Most auditors focus on Metal % (Fe/Au). However, Iron Ore is sold on Dry Metric Tonne (DMT).
  - *Risk:* If the lab reports 10% moisture instead of 8%, the company loses 2% revenue instantly.
  - Auditors must witness the 'Oven Drying' process. Ensure samples are weighed *immediately* after collection and *immediately* after drying. Delays cause evaporation, leading to understated moisture and customer penalties later."

D. The "105°C Rule"

The sample must be dried at 105°C +/- 5°C until it reaches a constant weight.

The Fraud: If a lab technician dries the sample for only 30 minutes instead of the required 2 hours, the sample remains partially wet.

- Result: The "Dry Weight" is artificially high.
- Calculation: The calculated Moisture % becomes lower (e.g., 8% instead of 12%).
- Financial Impact: The buyer pays for 4% water at the price of Iron Ore.

13.3 Detailed Audit Work Program: Quality Control (The Grade Audit)

Section 1: Sampling Integrity (The Input Risk)

Focus: Is the sample truly representative of the pile?

Sub Process	Red Flag / Risk	Audit Procedure (Field Steps)
1. The "Top Layer" Bias	Manual Sampling Fraud: When sampling trucks manually,	The "Dig Deep" Rule: Stop a manual sampling team. Instruct the loader to cut a trench

	laborers only pick from the top layer (which might be "Salted" with high-grade rocks), ignoring the low-grade material at the bottom.	through the stockpile. Take a sample from the Bottom 3rd of the pile. Compare this result with the official "Shift Sample."
2. Auto-Sampler Cutter	Mechanical Bias: The mechanical cutter on the conveyor belt does not cut the *entire* stream. It only cuts the middle (fines) and misses the edges (lumps), biasing the result.	Cutter Inspection: Inspect the Auto-Sampler while it is running. The cutter bucket must traverse the full width of the falling stream. If it stops halfway or misses the edges, the sample is non-representative.
3. Sample Tagging	Identity Swap: Samples are labeled with easily identifiable names like "High Grade Stack 4." This tells the chemist exactly which sample to "pass" and which to "fail."	Blind Coding Audit: Check the sample bags arriving at the lab. They should carry a Bar Code or a Random Number (e.g., #8021), NOT the location name. If the chemist knows the source, the process is compromised.
4. The "Cone & Quartering"	Preparation Bias: Using the old-fashioned "Cone and Quartering" method manually allows laborers to kick high-grade lumps into the "Save" pile and low-grade stones into the "Reject" pile.	Riffle Splitter Check: Ban manual quartering for high-value ore. Ensure a Riffle Splitter or Rotary Divider is used to reduce sample size. These machines ensure mathematically random splitting.

Section 2: Analytical Accuracy (The Lab Process)

Focus: Is the machine reading correctly?

Sub Process	Red Flag / Risk	Audit Procedure (Field Steps)
5. Calibration Drift	The "Happy Machine": The XRF machine has drifted and now reads everything 1% higher than reality. The chemist hasn't calibrated it in weeks.	The CRM Challenge: Walk into the lab with a Certified Reference Material (CRM)—a sachet of pre-tested powder with a known grade (e.g., 60.0% Fe). Ask the chemist to run it *immediately*. If the result is 61%, the machine is wrong.
6. Contamination (Salting)	Memory Effect: The pulverizer bowl is not cleaned after grinding a high-grade sample. The dust remains and "salts" the next low-grade sample.	The "Blank" Test: After the lab processes a high-grade sample (e.g., 65% Fe), ask them to run a "Blank" (Pure Silica Sand or Quartz). The Blank should read 0% Fe. If it reads 2% Fe, the bowl is contaminated.
7. Moisture Manipulation	The "Quick Dry" Fraud: Drying the sample for only 30 mins instead of 2 hours. The sample remains wet, inflating the "Dry Weight" and reducing the calculated Moisture %.	Temperature & Time Log: Check the "Oven Logbook." Does it show samples entering at 10:00 AM and leaving at 10:30 AM?
		* *Field Test:* Touch the sample immediately after weighing. If it

		is cool/damp, the moisture test is fake.
8. Sizing (Granulometry)	The "Lump as Fines" Scam: Selling "Lumps" (High Price) but classifying them as "Fines" (Low Price) to give a discount to the buyer.	Screen Test Verification: Take a bag of "Fines" ready for dispatch. Pour it over a 10mm Screen. If > 20% of the material stays on top of the screen (Oversize), the material is actually Lumps.

Section 3: Data Integrity & Governance

Focus: Preventing data tampering after the test.

Sub Process	Red Flag / Risk	Audit Procedure (Field Steps)
9. Sample Retention	Destroying Evidence: The lab throws away the "Pulp" (powder) immediately after testing so no one can re-check the result.	The "Umpire Library" Check: Visit the sample storage room. Are duplicate pulps from the last 3 months stored and indexed? Pick a random packet from last month and ask for a Re-Test. If the packet is missing, it's a major non-compliance.
10. Reporting Gap	Excel Manipulation: The XRF machine prints a receipt saying 61.2%, but the final Excel report sent to the customer says 62.1%.	Source Data Reconciliation: Select Sample Final Assay Reports. Ask to see the Raw Instrument Printout (or the un-editable internal machine log) for those specific Sample IDs. Match the numbers.
11. Umpire Variance	Collusion: The "Third Party Lab" always agrees with the Mine's	Variance Analysis: Compare the Internal Lab Results vs. Third Party Results for the last 50 shipments. If the Correlation

	internal lab, even when the buyer complains.	Coefficient is > 0.99 (Perfect Match), suspect collusion. A natural variance of ±0.2% is normal; a 0.0% variance is suspicious.

13.4 Field Insight: The "Blind Sample" Injection

The most powerful tool for an auditor in Chapter 13 is the Blind Sample Injection.

How to do it:

1. Create a Dummy: Take a sample of "Waste Rock" (0% Grade) or a "Standard" (Known Grade).
2. Disguise It: Put it in a regular sample bag. Label it "Production Sample- Shift B".
3. Insert It: Drop it into the sample receiving tray when the chemist isn't looking (or officially via the sample room).
4. Wait: Wait for the Daily Report.
5. The Trap:
    - If the report says "Grade: 62% Fe" for your Waste Rock sample, you know the lab is not testing at all—they are just "Cooking the Book" (Copy-pasting average grades).
    - This is conclusive proof of fraud.

13.4 Case Study: The "Salting" of the Core

In a famous gold mine fraud, the geologist added gold dust from his wedding ring (filed down) into the exploration drill core samples before sending them to the lab. The lab correctly analyzed high-grade gold. The "Chain of Custody" audit failed because the tampering happened before the sample reached the lab. The auditor caught it eventually by noticing the gold particles were "angular" (filed) under a microscope, rather than "rounded" (natural).

Part V: Supply Chain & Logistics (The Value Shield)

Chapter 14: Material Procurement

Coverage: Spares, Explosives, Fuel, Capex

Value Chain: Indenting (PR) > Tendering (RFQ) > Technical Eval > Commercial Neg. > PO Release > Delivery (GRN) > Payment

14.1 Introduction & Audit Objective

Procurement in the mining and heavy industry sector is distinct from general corporate procurement. It is characterized by High Value (Yellow Goods/Machinery), High Volume (Fuel/Explosives), and High Criticality (Spares).

For the Layman: The objective is ensuring we buy the right item, at the right price, right time to keep the mine running.

For the Expert Auditor: The objective is to dismantle the "Proprietary Myth" (paying a premium for branded generic parts), expose Cartelization (bid-rigging), and ensure Total Cost of Ownership (TCO) drives Capex decisions rather than just the lowest sticker price (L1).

Primary Audit Risks:

- Leakage: Buying standard industrial parts at "OEM" prices (300% markup).
- Cartelization: Vendors colluding to simulate competition.
- Inventory Bloat: Buying based on budget availability rather than consumption needs.
- Capex Mistakes: Buying cheap machinery that has high fuel consumption (Opex bleed).

14.2 Key Audit Focus Areas (Technical Concepts)

A. The "Proprietary Myth": OEM vs. OPM

This is the single largest area of cost leakage in heavy industry maintenance.

- The Concept: Mining OEMs (Original Equipment Manufacturers like CAT, Komatsu, Hitachi) often do not manufacture consumables like filters, bearings, hoses, or v-belts. They buy them from OPMs (Original Parts Manufacturers) like SKF, Parker, or Donaldson, box them in their branded packaging, and sell them at a massive markup.
- The Audit Strategy: Select 50 high-consumption "Proprietary" items. Trace the part number etched on the actual casting (not the box) against the OPM catalog.
- The Leakage Calculation: Leakage = (OEM Price - OPM Distributor Price) \ Annual Quantity

B. Total Cost of Ownership (TCO) in Capex

When buying a Dumper or Excavator, the "Acquisition Cost" is only 20% of the lifecycle cost. Fuel and Maintenance make up 80%.

- The Audit Check: Did the procurement committee select the L1 bidder (Lowest Price) or the L1 TCO bidder?
- TCO = {Acquisition Cost} + {Fuel Cost} + {Maint. Cost}- {Salvage Value}

(Ideally a discount rate formula should be used at least for all medium to high capex)

C. Fuel Procurement: Volume vs. Mass

Fuel is billed in Litres (Volume), but energy content is determined by Mass. Volume expands with heat.

- The Risk: If fuel is loaded at a hot terminal and delivered to a cool mine, the volume shrinks.
- Audit Requirement: Ensure payments are based on Density at converted rates.

14.3 Detailed Audit Work Program: Material Procurement

Section 1: Planning, Indenting & Demand Management

Focus: Optimizing what we buy, how we buy it, and preventing capital destruction.

Sub Process	Red Flag / Risk	Audit Procedure (Field Steps)
Synergy & Requirement Consolidation	Loss of Scale: Buying the same item (e.g., 15W40 Oil) piecemeal for Mine and Plant divisions, losing the "Bulk Discount" advantage.	Bulk PO Check: Check if separate indents were raised for the same material code by different departments in the same month/quarter/year. Did Procurement combine these into a single "Bulk PO" to negotiate a better rate?
Emergency Purchase Ratio	Wasteful Spending: High volume of "Emergency" or "Cash" purchases (>10%) indicating poor planning, leading to Air Freight or Rush Fee premiums.	Ratio Analysis: Calculate Value of Emergency POs / Total Procurement Value. If >10%, select top 10 emergency POs and verify if the urgency was genuine or caused by a forgotten indent.
"Min-Max" Logic Check	Inventory Bloat: Automatic re-ordering based on obsolete "Min-Max" levels (e.g., set for peak production) even when actual consumption has dropped by 50%.	Static vs. Dynamic Check: Review the ERP master data. Have the re-order levels been manually adjusted downwards to match the current lower production targets (e.g., during monsoon)?
Buy vs. Repair Decision	Capital Destruction: Buying new expensive assets (Motors, Gearboxes) when the existing ones could be	Repair Feasibility Audit: Select sample high-value POs for "Replacement Units." Ask for the "Beyond Economic Repair" (BER) certificate for the old unit. If it's

	repaired for 30% of the cost.	missing, the purchase was unjustified.
Buy vs. Rent Analysis	Opex Bleed: Paying monthly rental charges for heavy equipment (Cranes, Generators) that exceed the asset's capital cost over the project life.	NPV Model Review: For long-term rentals (>1 year), review the "Lease vs. Buy" financial model. Did Finance approve that renting is cheaper than buying based on Net Present Value?
Explosives Magazine Planning	Logistics Inefficiency: Ordering more explosives than the magazine can store, forcing vans to wait outside the gate (Demurrage).	Capacity Reconciliation: Compare PO Order Quantity vs. Available Magazine Capacity on the expected delivery date. If Order > Capacity, it is a planning failure.

Section 2: Tendering, Price Discovery & Fraud Detection

Focus: Ensuring fair competition and validating the "True" price.

Sub Process	Red Flag / Risk	Audit Procedure (Field Steps)
The "OEM vs. OPM" Check	Price Leakage (The Proprietary Myth): Paying a 300% markup for a "Branded Box" when the item inside is a standard industrial part available from the Original Parts Manufacturer.	Part Number Trace: Select 50 high-value consumables (Bearings, Filters). Trace the part number etched on the actual casting (not the box) against the OPM catalog (e.g., SKF, Parker).
Quote Leakage (H1 to L1)	Collusion: The Purchase Officer leaking the L1 price to	Bid History Forensic: Review the bid submission timestamps. Did a high bidder submit a "Revised

	a favored vendor (who was initially H1) to help them win the bid.	Offer" matching exactly just below the L1 price *after* the other bids were opened, without any change in scope?
Specification Consistency	Favoritism: Manipulating specifications to ensure a specific vendor becomes L1 artificially (e.g., asking Vendor A for "Standard Steel" and Vendor B for "High Tensile").	RFQ Comparison: Retrieve the "Request for Quote" (RFQ) email sent to all vendors. Ensure the technical specifications, delivery terms, and penalty clauses were identical for everyone.
Past Benchmarking	Price Creep: Vendors subtly increasing rates on "Repeat Orders" or claiming inflation when raw material indices are flat.	Last Purchase Price (LPP) Check: Compare the current L1 price with the price paid to the *same* vendor 6 months ago. The hike is without a corresponding rise in raw material indices?
Internal Benchmarking	Silo Buying: Different mine sites of the same company paying different prices for the same item (e.g., Tires, Explosives).	Group Price Parity: Compare the purchase price of common high-value items with the "Central Procurement" data or other site POs. Standardize rates across the group. (Also check for umbrella contracts)
The "Three-Quote" Forensic	Bid Rigging: The Purchase Manager producing 3 quotes to satisfy policy, but they are all from	Independence Check: Do not just count quotes. Look for: Same physical address, same font/formatting, or sequential invoice numbers (e.g., Ref-001

	friendly/related parties.	Ref-002) across different vendor quotes.
Landed Cost Analysis	Incorrect Evaluation: Selecting a vendor who is "L1 on Basic Price" but becomes "L3" when Freight, Duties, and Taxes are added.	TLC Verification: Ensure the Comparative Statement (CS) ranks vendors on Total Landed Cost (TLC). TLC = Basic + P&F + Freight + Insurance + Customs Duty + Non-Creditable GST.
Reverse Auction (e-Bidding)	Mock Auction: The auction is staged, with only one active bidder or a pre-agreed price drop at the last minute.	Time-Stamp Analysis: Check the auction log. If the L1 bidder dropped their price drastically (e.g., 10%) in the last 2 minutes without a counter-bid, the auction was likely pre-fixed.
Vendor Due Diligence (KYC)	Financial Risk: Awarding contracts to "Shell Companies" with no financial capacity, leading to project abandonment.	Capacity Check: For contracts > ₹1 Cr, check the vendor's Balance Sheet/Credit Rating. Identify odd vendors Eg: vendors with Net Worth < 10% of Contract Value.
Conflict of Interest (COI)	Fraud: Purchase Manager awarding contracts to companies owned by relatives.	Master Data Match: Run a fuzzy match between Vendor Master (Director Names/Addresses) and Employee Master (Emergency Contacts).

Section 3: Contract Terms, Quality & Execution

Focus: Getting what we paid for and preventing leakage.

Sub Process	Red Flag / Risk	Audit Procedure (Field Steps)

L1 Quality Adherence	"Cheap is Expensive" Trap: Buying sub-standard material to meet "Lowest Bidder" targets, leading to higher failure rates and downtime.	Failure Rate Analysis: Track the "Early Failure Rate" of the L1 vendor. If L1 goods fail twice as fast as L2, calculate the Total Cost of Ownership (TCO) to prove L1 is actually costlier.
Warranty Management	Financial Loss: Forgetting to claim free replacements for Capex components that fail within the warranty period.	Credit Note Search: Search for "Credit Notes" issued by vendors for warranty claims. Verify with Finance if these were actually utilized/deducted from the next payment.
VMI Payment Trigger	Cash Flow Loss: Paying for Consignment Stock (Vendor Managed Inventory) when it enters the gate, instead of when it is consumed.	Trigger Verification: Review the VMI contract. Ensure payment is triggered only upon "Withdrawal Slip" (Consumption), not upon "Gate Entry GRN."
Price Variation Clause (PVC)	Escalation Fraud: Vendors manipulating the formula or indices (e.g., Diesel/Steel price) to claim higher escalation than justified.	Formula Re-Check: Independently calculate the escalation amount. Ensure the "Base Index Date" matches the original Tender Date, not the Contract Signing Date.

Section 4: Peculiar Industry Risks (Mining Specific)

Focus: Specialized risks in Fuel, Heavy Machinery, and Imports.

Sub Process	Red Flag / Risk	Audit Procedure (Field Steps)

Fuel Procurement (Density)	"Phantom Volume" Loss: Paying for diesel volume expanded by heat (at 35°C) which shrinks when cooled, losing ~0.5% value instantly.	Density Conversion Check: Ensure PO terms specify payment based on "Density at 15°C" converted volume. Check if the "Dip Reading" volume was corrected using the ASTM conversion table.
Fuel Quality (Adulteration)	Engine Damage: Vendors mixing Kerosene/Solvents with Diesel. We pay for Diesel but get a mix that destroys injectors.	Lab Test: Send a random diesel sample from the bowser to an external lab. Test for Flash Point and Sulphur Content. Low flash point = Adulteration.
Scrap Buyback Clause	Asset Leakage: High-value scrap (OTR Tyres, Conveyor Belts) being stolen or sold cheaply because the replacement contract didn't cover it.	Net-Off Check: For replacement POs (e.g., New Belt), check if there is a line item for "Buyback of Old Item." If not, trace the old item to the Scrap Yard physically.
High Sea Sales (HSS)	Fund Diversion: Inflating the "Agency Commission" or "Loading" charges in a High Sea Sale structure to siphon funds abroad.	Title Transfer Audit: For imported machinery, verify the title transfer happened *before* the ship crossed the customs frontier. Audit the justification for any "Service Charges" added to the import value.
"Initial Spares" Dumping	Inventory Bloat: Vendors bundling a "2-Year Spares Package" with capital equipment	Usage Track: Review the "Initial Spares" bought with a machine 3 years ago. Are they still sitting in the box? If yes, flag this

	to dump slow-moving inventory.	"Bundling" as a restrictive trade practice.
GST Input Credit Classification	Statutory Non-Compliance: Misclassifying "Civil Foundations" (Ineligible Credit) as "Plant & Machinery" (Eligible Credit) to save 18% GST.	PO Review: Review POs for "Crusher Foundations" or "Silos." Confirm they are booked under the correct GL code (Civil Works) and that GST credit is not claimed.

14.4 Expert Insight: Key Frauds & Indicators

1. The "Split PO" Structuring

- The Fraud: The Plant Head has a signing authority of ₹5 Lakhs. A requirement of ₹20 Lakhs is split into four POs of ₹4.9 Lakhs each to bypass the CEO/Head Office approval.
- Red Flag: Multiple POs raised to the same vendor for similar material codes within a short window (e.g., 7 days).

2. Cover Bidding (Cartelization)

- The Fraud: The Purchase Manager wants to favor Vendor A. He asks Vendor A to arrange two "dummy quotes" from friendly companies (Vendor B and C) at a higher price.
- Red Flag:
    - Quotes from different vendors have identical PDF metadata (same "Author" or "Created" timestamp).
    - Sequential reference numbers (Vendor A: Ref-001, Vendor B: Ref-002).
    - Identical spelling mistakes in the "Terms & Conditions."

3. The "Consultant" Kickback

- The Fraud: A Capex vendor (e.g., a Crusher supplier) wins a contract. A separate "Consultancy Agreement" is signed between the Vendor and a shell company owned by the Purchase Manager's relative.
- Red Flag: High-value "Liaison" or "Commission" fees paid by the vendor to third parties, discovered during vendor due diligence.

14.5 Commonly Missed Aspects (The Grey Areas)

- Scrap Buyback Credits: When buying new Conveyor Belts or OTR Tyres, the contract should specify a "Buyback Price" for the old/scrap item. This is often missed, and the scrap is stolen or sold cheaply.
- Shelf-Life Expiry: Procurement often buys "bulk deals" on items like Resin capsules (for roof bolting) or Chemicals. If 30% of the stock expires before use, the "bulk discount" was a loss.
- Incoterms Violations: The PO says "FOR Destination" (Freight paid by vendor), but the Finance team mistakenly pays a separate Freight Bill submitted by the transporter.

14.6 The Indian Scenario: Key Challenges

1. High Sea Sales (HSS)

- Context: To save on Sales Tax (historically CST, now IGST cash flow), goods are sold while they are still on the ocean. Title transfers before the ship docks.
- Audit Risk: The "Loading" or "Agency Commission" added to the HSS value is often inflated to siphon funds abroad to a related party.

2. GST Input Tax Credit (ITC) Blockage

- Context: Section 17(5) of the CGST Act blocks credit for "Civil Works" or "Motor Vehicles" (in some cases).

- Audit Risk: Procurement often classifies "Plant Foundation Civil Work" as "Plant & Machinery" to claim GST credit. This invites massive litigation and penalties from tax authorities.

3. The "L1" Culture in PSUs

- Context: In India (especially PSUs), there is immense pressure to select the L1 (Lowest) bidder to avoid vigilance inquiries.
- Result: Companies buy sub-standard Chinese/local machinery that meets the *paper* spec but fails in the field.
- Audit Fix: Introduce strict "Quality-Cost Based Selection" (QCBS) where technical weightage is 70% and financial is 30%.

Value Chain: Scope Definition > Rate Analysis > Work Order (WO) > Mobilization > Execution (Measurement Book) > Bill Certification

15.1 Introduction & Audit Objective

Unlike buying a spare part (where you can count the inventory), buying a **Service** (Manpower, Transport, Excavation, Security) is buying an intangible. Once the day is over, the evidence of whether the security guard actually stood at the gate or slept is gone.

For the Layman: The objective is to ensure we pay for work actually done, not just for the "presence" of a contractor.

For the Expert Auditor: The objective is to validate "Proof of Performance" against Service Level Agreements (SLAs) and dismantle the mathematical manipulations hidden inside Price Variation Clauses (PVC).

Primary Audit Risks:

- Ghost Billing: Invoicing for manpower or trips that never happened.
- Escalation Fraud: Manipulating diesel/wage formulas to claim higher rates.
- Scope Creep: Charging extra for tasks that were already included in the base rate.
- Principal Employer Liability: The mine owner becoming liable for the contractor's unpaid Provident Fund (PF) or statutory dues.

15.2 Key Audit Focus Areas (Technical Concepts)

A. The Price Variation Clause (PVC) Math

In long-term contracts (3-10 years), rates cannot be fixed. They float based on Diesel prices and Minimum Wages.

- The Concept: The rate consists of a Fixed Component (Profit, Overhead) and a Variable Component (Fuel, Labor).
- The Audit Risk: If the contractor claims 100% of the rate is variable, they get a hike on their *profit* margin too when diesel rises. The Fixed component must remain fixed.

B. The MDO Model (Mine Developer & Operator)

Mining is increasingly outsourced to MDOs. They take the risk; the owner pays per tonne.

- The Risk: "Double Dipping." The MDO rate usually includes "All enabling infrastructure." However, during operations, the MDO often asks the mine owner to provide electricity, water, or road maintenance "temporarily." If the mine owner pays for these without deducting them from the MDO's per-tonne rate, it is double payment.

C. Service Level Agreements (SLAs) & Penalties

- The Concept: Payment should be linked to *output* (Performance), not just *input* (Manpower).
- The Audit Check: Review the "Penalty Register."
    - *Scenario:* The contract says "95% Dumper Availability required." Actual was 80%.
    - *Finding:* Was the penalty calculated and deducted? Or was it waived off citing "Force Majeure" or "Good Relations"?

D. The "Take-or-Pay" & Shortfall Penalties

MDO contracts usually mandate a "Minimum Guaranteed Production" (e.g., 90% of Annual Plan).

- The Revenue Loss: If the MDO produces only 70%, the mine loses the opportunity to sell that ore.
- The Fraud/Leakage: The MDO claims "Force Majeure" (e.g., Heavy Rains, Strike) to excuse the shortfall.

- Audit Procedure: Verify the "Rainfall Data" from the Indian Meteorological Department (IMD) against the MDO's claim. If the MDO claims "30 days lost due to rain" but IMD data shows only 10 rainy days, the Force Majeure claim is fraudulent, and the penalty must be deducted.

E. Internal Cost Estimate (The "Should Cost" Model)

- The Context: Unlike buying a bearing where you can check a catalogue price, services (like excavation or transport) are intangible and bespoke. Comparing "Quote A vs. Quote B" is dangerous if *both* vendors are overpriced.
- The Concept: The "Internal Estimate" (or Shadow Cost) is a zero-based cost calculation prepared by the Technical Department *before* tendering. It breaks down the service into fuel, wages, consumables, and margin.
- The Audit Check: Compare the Contract Award Price (L1) vs. the Internal Cost Estimate.
    - *Scenario:* The Internal Estimate for excavation was ₹50/Tonne. The L1 Bidder quoted ₹85/Tonne.
    - *Red Flag:* If the gap is >10-15%, it indicates either the Internal Estimate was flawed (poor planning) or the vendors have cartelized to quote high.
- Risk: Awarding contracts solely based on "L1" without an internal benchmark leads to paying a "Cartel Premium."

Example: Internal Cost Estimate (ICE) vs. Vendor Quote

The Scenario:

The Safety Department has raised a requisition to build a temporary scaffolding structure for painting a chimney/stack.

- Dimensions: 60 meters (Height) x 60 meters (Width).
- Duration: The structure will stand for 30 days.
- Scope: Erection, Rental of material for 30 days, and Dismantling.

Step 1: The Internal Cost Estimate (ICE) Calculation

(Prepared by the Mine's Civil Engineer or Quantity Surveyor before opening the bids)

The "Zero-Based" cost build-up logic:

Step 2: The Vendor Quote Comparison

(The Auditor verifies the ICE computation and assumption. Then compares the ICE against the bids received)

Step 3: The Audit Finding & Negotiation Strategy

Audit Query: "The L1 quote of ₹28.5 Lakhs is 45% higher than our Internal Technical Estimate of ₹19.6 Lakhs. There is no justification on file for awarding the contract at such a high variance."

Root Cause Investigation: Upon questioning, the vendors claimed "Height Risk Premium" because working at 60 meters requires specialized "Height Pass" labor which costs double the standard rate.

Corrective Action (Negotiation): The Auditor/Procurement team recalculates the ICE with double labor rates (₹2,000/day).

- Revised ICE: ₹22,50,000.
- Result: The L1 quote of ₹28.50 Lakhs is *still* too high.
- Outcome: Procurement negotiates Vendor A down to ₹23.00 Lakhs, saving the company ₹5.5 Lakhs instantly.

15.3 Detailed Audit Work Program: Service Procurement

Section 1: Planning, Estimation & Scope Definition

Focus: Defining the "True" requirement and cost before going to market.

Sub Process	Red Flag / Risk	Audit Procedure (Field Steps)
Internal Cost Estimate (ICE)	The "Cartel Premium": Awarding a contract to L1 without realizing that *all* bidders have cartelized and quoted	The "Should Cost" Verification: Obtain the internal "Zero-Based" cost estimate prepared by the technical team *before* the tender. Compare ICE vs. L1 Price.

	50% above the true market cost.	Test: If L1 is >15% higher than ICE, demand a "Price Justification" or re-tender.
In-House vs. Outsource	Operational Inefficiency: Outsourcing core activities (e.g., Maintenance) that could be done cheaper by existing idle internal staff.	Cost-Benefit Analysis: Check the justification note. Did the department compare the "Contractor Rate" vs. the "Marginal Cost" of using internal labor + overtime?
Scope Duplication (MDO)	"Double Dipping": The Mine Developer & Operator (MDO) rate includes "Enabling Infrastructure," yet the Mine Owner awards separate contracts for Road Maintenance or Lighting.	Scope Exclusion Check: Read the "General Conditions of Contract" (GCC) in the MDO agreement. List all inclusions. Cross-check against the Service PO Register. Any overlap (e.g., Water Sprinkling) is a duplicate payment.
Synergy & Consolidation	Fragmented Buying: Hiring different Security/Housekeeping agencies for the Mine, Plant, and Colony at different rates.	Group Rate Check: Compare the "Man-Day Rates" for unskilled labor across all 3 zones. If the Mine pays ₹800 and the Plant pays ₹700 for the same guard, recommend a consolidated "Group Contract."

Section 2: Tendering, Rates & Vendor Selection

Focus: Ensuring the price is genuine and the process is fair.

Sub Process	Red Flag / Risk	Audit Procedure (Field Steps)

Specification Consistency	Manipulated Bids: Vendor A quotes for "24-Hour Coverage" while Vendor B quotes for "12-Hour Coverage" because the scope sent to them was different.	RFQ Audit: Retrieve the email/RFQ sent to all bidders. Did they receive identical "Scope of Work" (SOW) and "Service Level Agreements" (SLA)? If specs differ, the L1 status is fake.
Quote Leakage (H1 to L1)	Collusion: The Purchase Officer leaks the L1 price to a preferred vendor (who was initially H1) to help them revise their bid and win.	Timestamp Forensic: Review the bid history. Did a high bidder submit a "Revised Offer" matching exactly just below the L1 price *after* other bids were opened, with no change in scope?
Internal Benchmarking	Rate Disparity: Paying higher rates for "Civil Excavation" than what was paid in the previous year's contract for the same soil type.	Rate Trend Analysis: Compare the unit rates (per m^3 or per km) with the Last Purchase Price (LPP). If the rate hiked by 20% but Diesel/Labor indices moved only 5%, the hike is unjustified.
External Benchmarking	Market Alignment: Paying significantly above the regional market rate for standard services (e.g., Bus Hiring, Security).	Regional Check: Call a local vendor (anonymously) or check with a neighboring industry/mine. If the market rate for a "50-Seater Bus" is ₹60k/month and you pay ₹90k, investigate.
Mobilization Advance	Financial Risk: Releasing a 10% advance to a	BG Verification: Ensure no Mobilization Advance is paid without a confirmed Bank

	financially weak contractor who uses it to pay old debts and then abandons the site.	Guarantee (BG). Check the BG expiry date—it must cover the project duration + claim period.
Vendor Due Diligence (KYC)	Performance Risk: Awarding critical service contracts to "Shell Companies" with no assets or manpower.	Capacity Check: For contracts > ₹1 Cr, review the vendor's Balance Sheet. Do not award if Net Worth is < 10% of Contract Value.
Conflict of Interest (COI)	Fraud: Purchase Manager awarding service contracts to companies owned by relatives.	Master Data Match: Run a fuzzy match between Vendor Master (Director Names) and Employee Master (Emergency Contacts).

Section 3: Execution, Measurement & Billing

Focus: Did they actually do the work? (The "Invisible Product" Risk).

Sub Process	Red Flag / Risk	Audit Procedure (Field Steps)
PF vs. Invoice Reconciliation	Ghost Workers / Fake Billing: The contractor bills for 100 workers but deposits PF for only 80. The variance (20 heads) represents fake billing or illegal employment.	Headcount Triangle: Compare three documents: 1. Invoice Quantity (Man-days billed). 2. Gate Attendance Register (Physical Entry). 3. PF ECR Statement (Statutory Deposit). Test: If Invoice > PF Record, disallow the difference immediately.

Measurement Book (MB)	The "Hidden Work" Fraud: Civil contractors claiming "Hard Rock Excavation" (expensive) for what was actually "Soft Soil" (cheap).	Visual & Blasting Check: Visit the site. If they claimed Hard Rock (>500/m3), check the Explosives Consumption Log. No blasting = No hard rock. Also, check "Pre-Levels" and "Post-Levels" in the survey software.
Vehicle Hiring (Logbooks)	The "Odometer" Fraud: Hired vehicles logging 300km/day while parked at the gate. Drivers spinning odometers or logging fake trips.	GPS/FastTag Reconciliation: Pick 5 random days. Compare Logbook Entries vs. GPS Tracking Reports or FastTag Toll Deductions. If the log says "Airport Trip" but FastTag shows no toll paid, the trip is fake.
Manpower vs. Output	Headcount Padding: Contracts based on "Number of Persons" rather than "Output." The contractor supplies 50 lazy workers instead of 30 efficient ones to maximize billing.	SLA Validation: Shift contracts to "Output-Based" (e.g., Clean 500 m2 area) instead of "Manpower-Based." If Manpower-based, conduct a Surprise Headcount at various timings of the shift.
AMC Visits (Ghost Visits)	Service Default: Paying for 4 quarterly visits/year when the engineer visited only once.	Gate Entry Correlation: Select 5 "Service Reports" attached to the invoice. Check the Main Gate Visitor Log or Security System. Did the specific technician actually enter the plant on that date?

L1 Quality & SLA Penalty	Performance Failure: The L1 vendor fails to meet targets (e.g., 90% Dumper Availability) but is paid the full amount.	Penalty Deduction Check: Review the "Penalty Clause" in the contract. Calculate the actual penalty based on performance data. Has this amount been deducted from the bill? If waived, look for the "Waiver Approval."

Section 4: Mining Specific & Statutory Risks

Focus: High-value operational and legal risks.

Sub Process	Red Flag / Risk	Audit Procedure (Field Steps)
Overburden (OB) Lead	Distance Inflation: Paying transport contractors for a 4km "Lead" (Haul Distance) when they are dumping at a 2km yard.	Odometer Test: Physically ride in a dumper from the Face to the Dump Yard. Record the reading. If the actual distance is 2.2km but the contract pays for the "3-4km Slab," it is a massive overpayment.
MDO Shortfall Penalty	Revenue Loss: The MDO fails to produce the "Minimum Guaranteed Tonnage" but claims "Force Majeure" (Rain) to avoid penalties.	Weather Data Forensic: Compare the MDO's "Rain Day" claim with official IMD (Met Dept) data. If IMD says "Sunny" but MDO claims "Flooding," impose the penalty.
Crushing Yield Loss	Double Billing: The contractor is paid on "Feed Tonnage." They rush material,	Recirculation Check: Ensure payment is based on "Final Product Output", not Feed. Check the "Recirculation Load" log; high

	creating high "Oversize" rejection, which they re-feed and bill again.	recirculation means process inefficiency.
Price Variation Clause (PVC)	Escalation Theft: Contractor claiming escalation on the *Gross Bill Value*, ignoring the Fixed Component.	Formula Re-Calc: Ensure escalation is applied *only* to the variable components (Fuel/Labor) as per the specific formula in the contract. Verify the Base Index Date matches the Tender Date.
PF/ESI Compliance	Liability Risk: Contractor collecting PF from workers but not depositing it. The Principal Employer (Mine) is liable.	ECR Verification: Do not accept challan copies. Demand the Electronic Challan Return (ECR). Match 5 random names from the site attendance register to the ECR list to ensure coverage.
GST Reverse Charge (RCM)	Tax Leakage: Paying GST to a Goods Transport Agency (GTA) or Legal Firm, where the liability is on the Mine (Recipient).	Ledger Scrutiny: Filter Service POs for GTA/Legal/Sponsorship. Ensure no GST was paid to the vendor, and the liability was booked under RCM Payable.
Safety Compliance	Accident Risk: Contractor using uncertified lifting tools or untrained drivers.	T&P Audit: Inspect the contractor's Tools & Plant (T&P) Register. Are Cranes certified (Form 11)? Check Operator Licenses for heavy machinery.

15.4 Expert Insight: Key Frauds & Indicators

1. The "Base Date" Manipulation (PVC Fraud)

- The Mechanism: The contractor negotiates to set the "Base Diesel Price" index artificially low in the contract terms.
    - *Example:* Current Price = ₹90. Contract Base Price set at = ₹80.
    - *Result:* On Day 1 of the contract, the contractor claims an escalation bill because ₹90 > ₹80. They generate immediate profit without doing work.
- Red Flag: An "Escalation Bill" submitted along with the very first "Running Account (RA) Bill."

2. The "Negative Scope" Trick

- The Mechanism: The original tender included "Haul Road Dust Suppression" (Water sprinkling). After winning the contract, the contractor claims their water tanker is broken. The Mine Owner uses their own tanker to suppress dust to avoid pollution fines.
- The Fraud: The contractor continues to bill the full rate per tonne, *including* the cost of dust suppression, even though the Mine Owner is doing the work.
- Red Flag: Mine Owner's equipment logbooks showing work in areas assigned to the Contractor.

3. The "Ghost Shift"

- The Mechanism: In 24x7 operations, contractors bill for 3 shifts (8 hours each).
- The Fraud: They actually run 2 shifts of 12 hours each.
    - *Billed:* 30 men (10 per shift x 3).
    - *Actual:* 20 men (10 per shift x 2).
    - *Leakage:* 10 men's wages stolen.
- Red Flag: Biometric logs showing "Punch In" at 8 AM and "Punch Out" at 8 PM (12 hours) consistently.

4. The "Ghost Shift" reverse

The Mechanism: In emergency / shut down period 24x7 operations, contractors were hired for 2 shifts (12 hours each), to reduce wastage time in shift changeover.

The Fraud: They actually run 3 shifts of 8 hours each.

1. *Billed:* 40 men wages (10 men x 2 shift X 2 for Overtime).
2. *Actual:* 30 men wages (10 per shift x 3).
3. *Leakage:* 10 men's wages stolen, and also shift changeover time wasted against the planned, increasing total time required.

15.5 Commonly Missed Aspects (The Grey Areas)

- Safety Violations as Revenue: Finance teams rarely talk to Safety Officers. If a contractor's staff violated safety norms (e.g., not wearing PPE), the contract usually allows for a fine (e.g., ₹5,000 per instance). These fines are rarely deducted from the invoice because the "Safety Challan" never reaches the Accounts department.
- The "De-mobilization" Clause: Contracts often state that the Mobilization Advance is for bringing machinery *to* the site. If the contractor leaves early or reduces the fleet, a pro-rata portion of the mobilization cost should be clawed back. This is almost never done.

15.6 The Indian Scenario: Key Challenges

1. Principal Employer Liability (The PF Trap)

- Context: Under Indian Law (EPF & MP Act), if the Contractor fails to deposit the Provident Fund (PF) of their workers, the Liability falls on the Principal Employer (The Mine Owner).

- Audit Check: Do not just accept the "PF Challan" copy attached to the invoice. Use the TRRN (Temporary Return Reference Number) to verify on the EPFO portal that the money was *actually credited* to the specific worker names deployed at your site.
- Fraud: Contractors collecting PF from workers, generating a fake Challan, and vanishing.

2. The "Liaison" Consultant

- Context: In India, land acquisition and local village relations are difficult. Companies hire "Liaison Consultants."
- Audit Risk: These contracts are often vague ("Maintain good relations"). They are frequently vehicles for siphoning cash for bribes or speed money.
- Audit Control: Demand specific "Activity Reports" (e.g., Minutes of Meeting with Village Sarpanch). "General Expenses" should be rejected.

3. Cartelization in Local Transport

- Context: Local truck unions in mining districts (e.g., Odisha, Jharkhand) often fix rates and refuse to let outside transporters bid.
- Audit Reality: The "Three Quote" rule fails here.
- Audit Strategy: Benchmark the rate against the "District Schedule of Rates" (DSR) published by the local government, rather than trying to find fake competitive quotes.

Chapter 16: Warehouse and Stores Management

This chapter shifts focus from *buying* (Procurement) to *holding* (Warehousing). In mining, the warehouse is often an open yard worth crore of rupees, making it a high-risk area for physical deterioration and pilferage. This area is high-risk in mining because millions of dollars of capital spares (Motors, Tyres, Bearings) sit in inventory.

Value Chain: Receipt (GRN) > Binning/Storage > Reservation > Issue (Slip) > Return (Reverse Logistics) > Scrap Disposal

16.1 Introduction & Audit Objective

The Store is the "Bank Vault" of the mine, but instead of cash, it holds Bearings, Conveyor Belts, Motors, and Tyres.

- For the Layman: The goal is to ensure that what the computer says we have (Book Stock) matches what is physically in the bin (Physical Stock).
- For the Expert Auditor: The goal is to prevent Capital Erosion through Obsolescence (rust/expiry) and to detect "Shadow Inventory" (hoarding). The auditor must verify not just Quantity, but Quality and Value.

Primary Audit Risks: The primary risks are Theft, Obsolescence (Rusting), and Fake Consumption.

- Phantom Stock: The ERP shows 10 items, the bin is empty (Theft/Data Error).
- Obsolescence: Holding high-value parts for machines that were scrapped 5 years ago.
- Preservation Failure: Expensive assets rotting in the open yard due to poor storage.
- Scrap Leakage: New parts issued without receiving the old/damaged part (The "New-for-Old" Fraud).

16.2 Key Audit Focus Areas (Technical Concepts)

A. Inventory Classification: ABC & XYZ Analysis

You cannot audit every nut and bolt with the same rigor. You must segment the inventory.

- ABC Analysis (Consumption Value):
  - A-Class: High Consumption Value (e.g., Diesel, Explosives, Tyres). *Strict Control.*
  - C-Class: Low Consumption Value (e.g., Washers, Cotton waste). *Open Bin.*
- XYZ Analysis (Closing Stock Value):
  - X-Class: High Value sitting in stock.
  - Z-Class: Low Value sitting in stock.
- The Audit Sweet Spot: The "X" Category (High Value, Slow Moving). This is where "Dead Stock" hides.

B. Preservation Standards (The "Rot" Factor)

Mining stores are often dusty, hot, and exposed.

- Conveyor Belts: Must be stored in "Cradles" or upright. If stored flat on their side (edges) in the mud, moisture wicks into the fabric ply, destroying the belt before it is ever used.
- Bearings: Must be stored flat and rotated periodically to prevent "Brinelling" (permanent indentation of the race).
- O-Rings/Hydraulics: Must be kept away from sunlight (UV radiation causes rubber to crack).

C. The "Bone Yard" (Scrap Management)

The Scrap Yard is the "Shadow" of the Store.

- The Rule: For every "Capital Spare" (Motor, Pump, Cylinder) issued, a corresponding "Old/Damaged" item must be returned to the Scrap Yard.

- The Audit: If the Store issued 10 new Hydraulic Pumps this month, are there 10 old pumps in the Quarantine/Scrap area? If not, the new pumps were likely sold or stolen.

D. The "Refurbished Spare" Loophole- The Fraud Scenario (The "Phantom Swap"):

1. The Trigger: A breakdown occurs. The Maintenance Engineer requests a Brand New Valve (Value: ₹50,000) from the Store.
2. The Swap: In the workshop, the engineer has an old, locally repaired (refurbished) valve. He installs this Refurbished Valve into the machine.
3. The Theft: The machine starts working perfectly. The Engineer then takes the Brand New Valve (still in the box) and sells it or takes it home.
4. The Cover-Up: Since the machine is running, production is happy, and no one suspects that a used part was used instead of the new one.

The Audit Control: "One-to-One" Exchange Policy: To stop this, the auditor must enforce a strict "Old Part Return" logic. The store must not issue a new item unless the old/failed item is physically returned or accounted for.

16.3 Detailed Audit Work Program: Warehouse & Stores

Section 1: Physical Stock Verification & Accuracy

Focus: Does the physical stock match the ERP system?

Sub Process	Red Flag / Risk	Audit Procedure (Field Steps)
Perpetual Inventory Count (PIC)	The "Adjusted" Count: Storekeepers manually "adjusting" stock variances in the system without approval to	The "Blind Count" Test: Do not give the Storekeeper the "System Stock Sheet." Give them a blank sheet and ask them to count 10 high-value items (e.g., OTR Tyres, Cables). Compare their count vs.

	hide shortages before the annual audit.	SAP System Stock. Any variance is a potential theft.
Weighbridge vs. Dip Check	Bulk Material Fraud: For bulk items like Coal/Diesel/Explosives, relying on "Book Stock" instead of actual volumetric measurement.	Volumetric Verification: For Diesel tanks, perform a Dip Check and convert to Liters using the calibration chart. For Coal/Ore stockpiles, use a Drone Survey or Total Station survey to calculate volume. Compare "Calculated Volume" vs. "Book Stock."
Binning Accuracy	Location Chaos: Parts stored in the wrong bins intentionally so they "cannot be found" during issue, forcing the company to buy new ones (while the old ones are stolen later).	Bin Audit: Pick 10 items from the SAP system and go to the designated "Bin Location." Are they there? *Conversely:* Pick 10 items physically from the shelf and check if they are correctly updated in the SAP Bin Master.

Section 2: Goods Receipt & Issue (GRN & GIN)

Focus: Preventing fake receipts and unauthorized issues.

Sub Process	Red Flag / Risk	Audit Procedure (Field Steps)
The "Blind" GRN	Short Supply: The Storekeeper creating a Goods Receipt Note (GRN) for 100 items based on the Vendor's Invoice, without actually counting the boxes (which contain only 80).	Staging Area Check: Visit the "Inward Bay." Check a newly arrived consignment. Ask to see the "Tally Sheet" (rough handwritten count) prepared by the labor. If there is no Tally Sheet, the GRN was likely created blindly from the invoice.

Consumption vs. Issue	The "Hoarding" Risk: Production departments issuing spares (e.g., 50 Valves) "for stock" at their site cupboards, charging it to consumption immediately. This hides true inventory levels.	Sub-Store Raid: Visit the Maintenance/Electrical workshop cupboards. If you find new, unused bearings/valves that were "Issued" 3 months ago, it is "Fake Consumption." These items should be returned to the Main Store.
FIFO Violation	Obsolescence Risk: Issuing new batches of chemicals/oil while old batches expire and rot at the back of the shelf.	Expiry Date Check: Check items with shelf life (Cement, Resin, Paint). Are the "Oldest Batches" stored at the front? If a 2024 batch is in front of a 2023 batch, FIFO (First-In-First-Out) is violated.

Section 3: Preservation & Housekeeping

Focus: Protecting the asset value.

Sub Process	Red Flag / Risk	Audit Procedure (Field Steps)
Outdoor Storage	Asset Deterioration: Storing sensitive items (Conveyor Belts, Motors) in the open yard due to laziness. Sunlight destroys rubber; rain destroys insulation.	Yard Walkthrough: Inspect the Open Yard. Are Conveyor Belts covered with tarpaulins? Are Motors kept on raised pallets to prevent water logging? If a ₹10 Lakh motor is rusting in mud, write a major non-compliance.
Kitting & Cannibalization	The "Skeleton" Asset: Taking parts (e.g., a fan, a fuse) out of a new machine in stock	Visual Inspection: Inspect high-value "Insurance Spares" (e.g., Spare Engine). Is it sealed? Open the panel—are parts missing?

	to fix an old machine, leaving the new machine useless.	"Cannibalization" is often undocumented, leading to surprises when the spare is actually needed.

Section 4: Obsolescence & Non-Moving Stock

Focus: Cleaning up the books.

Sub Process	Red Flag / Risk	Audit Procedure (Field Steps)
Non-Moving Analysis	Capital Blockage: Holding 5 years' worth of inventory for a machine that was scrapped last year.	Zero-Consumption Filter: Extract the "Stock Ageing Report." Filter for items with "Last Issue Date" > 3 Years. *Field Step:* Ask Maintenance: "Do we still own the machine this part belongs to?" If no, the stock must be written off and sold as scrap immediately.
Duplication of Codes	Inventory Bloat: The same bearing exists under two different Material Codes (e.g., "Bearing 6205" and "Ball Brg 6205"). This causes the system to order new stock even when stock exists under the other code.	Description Search: Export the Material Master. Sort by Description. Look for near-duplicates. *Test:* If Code A has 0 stock and Code B has 10, but the Purchase Request was raised for Code A, it is a Master Data failure.

Section 5: Returnable Gate Pass (RGP) & Scrap

Focus: Leakage at the gate.

Sub Process	Red Flag / Risk	Audit Procedure (Field Steps)

RGP Pending > 6 Months	Asset Leakage: A motor sent outside for "Rewinding/Repair" 12 months ago has not returned. It is likely lost or stolen by the vendor.	RGP Aging Report: Filter the "Open RGP Register" for items pending > 6 months. Call the vendor immediately. If they say "It was scrapped," ask why the Scrap Value wasn't recovered.
Scrap Weight Reconciliation	Theft: The scrap truck weighs 10 Tonnes at the Weighbridge, but the Gate Pass says 8 Tonnes. The 2 Tonne difference is stolen metal.	Cross-Check: Compare the "Gate Pass Estimated Weight" vs. "Actual Weighbridge Slip". The invoice to the scrap buyer must be based on the *Weighbridge* weight, not the Gate Pass estimate.

Section 6: Refurbished Inventory Control

Sub Process	Red Flag / Risk	Audit Procedure (Field Steps)
The "Old for New" Reconciliation	Asset Leakage: Issuing 10 new motors but receiving only 8 old motors in the scrap yard. The variance (2) implies that 2 new motors were stolen, or 2 old motors were illegally refurbished and reused.	The "Mass Balance" Test: 1. Select a category (e.g., Hydraulic Pumps). 2. Check Total New Issued in last 6 months (e.g., 50 units). 3. Check Total Old Returned to Scrap Yard + Repair Bay (e.g., 40 units). 4. Variance (10 Units): Where are they? If they are not in the machine, the new spares were likely stolen.
Refurbished Stock Visibility	Hidden Inventory: Maintenance teams keeping a "Private Store" of refurbished	Workshop Raid: Visit the Maintenance Shop floor. Open the cupboards. If you find "Ready to Use" spares that are not

	parts inside their workshop cupboards, invisible to SAP.	tagged in SAP, this is a fraud risk. These items must be moved to the Main Store and valued at 50% of New Price.
Repair vs. Scrap Decision	Value Destruction: Sending a repairable motor to the Scrap Yard (sold for ₹50/kg) instead of the Refurbishment Section. The scrap buyer pays the scrap rate but refurbishes it and sells it back to the company as "New."	Scrap Bin Inspection: Inspect the "Metal Scrap" bin before it leaves the gate. Are there repairable items (Valves/Motors) inside? Test: Pick one item, ask the workshop: "Why was this declared BER (Beyond Economic Repair)?" If there is no technical report, it's a leakage.
Life Cycle Tracking	Premature Failure: A refurbished part is installed but treated as "New" in the system. It fails in 3 months (instead of 2 years).	Serial Number Track: Does the ERP track the Serial Number of the refurbished item? Test: If a "New" issue fails in 3 months, check if the physical serial number matches a previously repaired unit. This proves the "Phantom Swap."

Section 6 – Project & Commissioning Spares (Capex)

Focus: Project Spares that arrives with the machine but often vanishes. These are high-value spares supplied along with new machines (Capex) that often bypass the warehouse controls because they are delivered directly to the project site.

Sub Process	Red Flag / Risk	Audit Procedure (Field Steps)

The "Packing List" Reconciliation	"Ghost Inventory": The vendor sends a "Start-up Kit" (Filters, Gaskets, Sensors) inside the main machine crate. The Project Manager keeps it in the site office cupboard. It never enters SAP. *Result:* The Store buys new filters next year because they don't know these exist.	Logistics vs. Store Match: 1. Obtain the Original OEM Packing List or Bill of Entry from the Logistics/Import department (not the Store). 2. Compare it line-by-line with the Store GRN. Test: If the Packing List shows 50 loose items but the GRN shows only 1 line item ("Machine Lot"), the spares are un-accounted for.
Warranty Cannibalization	The "Free Fix" Loss: The new machine fails during the warranty period. To save time, the Project Manager uses the "Commissioning Spare" to fix it, instead of claiming a free replacement from the vendor.	Replenishment Audit: Track the consumption of Project Spares during the first 12 months (Warranty Period). Test: For every spare consumed, is there a corresponding "Free of Cost (FOC) Replenishment" from the vendor? If not, the company effectively paid for a warranty defect.
Capitalization vs. Inventory	Accounting Mismatch: "Insurance Spares" (e.g., a large Spare Motor) are often capitalized with the main machine (Depreciated over 15	Asset Tag Check: Identify high-value spares lying in the project yard. Ask Finance: "Are these booked as Inventory (Current Asset) or Fixed Asset?" Test: If booked as Fixed Asset, physically verify they are tagged.

	years). If they are consumed in Year 2, the asset value remains inflated.	If one is missing (consumed), check if a "Partial Asset Write-Off" entry was passed.
Master Data Duplication	Buying What We Have: The Project team has a spare "Hydraulic Pump" sitting in a crate without a Material Code. The Maintenance team, unaware of this, raises a PR to buy the same pump.	Un-coded Stock Survey: Visit the Project "Lay-down Area." Look for crates marked "Spares." Test: Open a crate. Pick a part number. Search it in SAP. If the system shows "No Record," the company is at risk of buying duplicate stock.

16.4 Expert Insight: Key Frauds & Indicators

1. The "New for Old" Swap (The Mechanic's Scam)

- The Mechanism: A mechanic requests a new Self Starter (INR 500). He draws it from the store.
- The Fraud: He cleans up the old Self Starter and keeps it on the machine. He takes the *new* unit and sells it in the local grey market. Or, he installs the new one and sells the old one (which was repairable) as scrap.
- Red Flag: High consumption of "Rotable" items (Alternators, Starters, Pumps) without a corresponding increase in fleet hours.

2. The "Virtual Stock" Adjustment

- The Mechanism: The Store Manager knows 50 items are missing (stolen).

- The Fraud: Right before the Physical Audit, he passes a "Stock Adjustment" entry in SAP, or issues the missing items to a dummy "Cost Center" (e.g., "General Consumption").
- Red Flag: A spike in "Consumption" or "Write-off" entries in the last week of the month/quarter.

3. The "Shadow Inventory" Hoarding

- The Mechanism: Operations teams don't trust the Store. When they indent for 10 bearings, they order 20.
- The Fraud: They use 10 and hide 10 in their personal cupboards in the workshop. This "Shadow Inventory" is expensed off the books but physically exists. It is highly prone to theft.
- Red Flag: Finding operational spares stored in unauthorized locations (Supervisors' cabins, electrical substations).

16.5 Commonly Missed Aspects (The Grey Areas)

- The "Emergency Purchase" Loophole: Items bought on "Emergency Cash Purchase" often bypass the Store completely. They are bought and directly given to the user. These are never recorded in the Stock Register, making them easy to steal or duplicate billing.
- Diesel Dispersion: In mines, Diesel is often dispensed to "Mobile Service Vans" (Bowsers). The Store treats the Bowser as a "Consumer" and expenses the fuel. The Audit must treat the Bowser as a "Sub-Store"—the fuel isn't consumed until the Bowser dispenses it to a Dumper. Also check Bowser Calibration: Pour 20 Litres from the Bowser nozzle into a "Certified Measure Can." If the meter reads 20L but the can is only 18L full, the nozzle is rigged to short-fuel the dumpers.
- Warranty Claims on Scrap: When a battery fails within 1 year, it goes to the scrap yard. The Store often sells it as lead scrap instead of claiming the Warranty Replacement.

16.6 The Indian Scenario: Key Challenges

1. "Jugaad" Storage & The Open Yard

- Context: Indian mines often lack covered warehousing for heavy items.
- Audit Find: Expensive conveyor belts (worth Crores) stored vertically on their edges in waterlogged yards. The "Chalta Hai" (it's okay) attitude towards preservation leads to massive capital destruction.
- Impact: When the belt is finally installed, it fails in 2 months due to "Ply Separation."

2. GST Input Tax Credit (ITC) Reversal

- Context: Section 17(5)(h) of the CGST Act (India) states that ITC must be reversed for goods *lost, stolen, destroyed, or written off*.
- Audit Risk: If the store writes off ₹1 Crore of obsolete stock, the Finance team must reverse ₹18 Lakhs of GST credit previously claimed. This is frequently missed, leading to tax penalties + interest.

3. The "Local Mafia" in Scrap Disposal

- Context: Scrap disposal in mining belts (Jharkhand/Odisha/Chhattisgarh) is often controlled by local strongmen.
- Audit Risk: Weighbridge manipulation during scrap sale. The truck weighs 15 Tonnes empty (Tare) and 25 Tonnes loaded. The Mafia forces the weighbridge operator to record the loaded weight as 20 Tonnes.
- Audit Control: Surprise cross-verification of Scrap Trucks at a public weighbridge 50km away from the mine.

Chapter 17: Inward, Intra-plant, and Inter-plant Logistics

Chapter 17 focuses on "Cost" logistics (bringing materials in and moving them around), while Chapter 18 focuses on "Revenue" logistics (getting the product to the customer).

Value Chain: Gate In > Weighment (Gross) > Unloading Point > Weighment (Tare) > Quality Sampling > Gate Out

17.1 Introduction & Audit Objective

Logistics is the "artery" system of the mine. It connects the Purchase Order to the Inventory (Inward), the Pit to the Plant (Intra), and the Mine to the Power Plant/Washery (Inter).

For the Layman: The goal is to ensure that "What was sent is what arrived" (Quantity) and "It arrived efficiently" (Time/Cost).

For the Expert Auditor: The goal is to detect "Phantom Trips" (Intra-plant fraud), "Transit Loss Arbitrage" (Inter-plant theft), and "Tare Weight Manipulation" at the weighbridge.

Primary Audit Risks:

- Inward: Suppliers "wetting" coal/mineral to sell water as commodity.
- Intra-plant: Contractors billing for "Double Handling" (moving the same pile twice).
- Inter-plant: High-grade ore leaving Unit A but Low-grade ore arriving at Unit B (Theft en route).
- Weighbridge Fraud: Tampering with sensors to under-weigh (sales) or over-weigh (purchases).

17.2 Key Audit Focus Areas (Technical Concepts)

A. The Weighbridge: The "Zero" Error

The weighbridge is the single most critical control point.

- Tare Weight Manipulation: Net Weight = Gross (Loaded) - Tare (Empty).
    - *The Fraud:* A truck enters with a hidden water tank full (Heavy Tare). It dumps the water inside the plant, loads the material, and weighs out. The system subtracts the *Heavy* Tare from the *Actual* Gross, resulting in a lower Net Weight.
- Audit Check: Analyze the "Tare Variance Report." Does the same truck ID have a Tare weight of 12T on Monday and 14T on Tuesday? A 2-tonne variance is impossible (fuel is only ~300kg).

B. Intra-Plant Logistics (The "Trip Count" Trap)

Moving ore from the Pit Head to the Crusher (2km distance) is often paid on a "Per Trip" or "Per Tonne" basis. The Risk: "Phantom Trips." The contractor runs the truck in a loop past the RFID sensor without actually loading material, or loads the bucket only 50% full to increase trip count.

C. Inter-Plant Logistics (Transit Loss)

When moving coal from a Captive Mine to a Power Plant (50km away), some loss occurs (wind, moisture).

- The Norm: Industry standard transit loss is 0.5%- 0.8%.
- The Leakage: If the loss is consistently 0.2% (too good to be true), the weighbridges are rigged. If it is 2.0%, coal is being stolen.

17.3 Detailed Audit Work Program: Logistics & Transport

Logistics in mining is a "High Velocity, High Volume" area. The primary risks are Transit Theft (Pilferage), Fake Trips (Ghost Billing), and Diesel/Freight Rate Manipulation.

Section 1: Trip Validation & Route Integrity

Focus: Did the trip actually happen, and did it follow the authorized route?

Sub Process	Red Flag / Risk	Audit Procedure (Field Steps)
GPS Geofence Audit	Route Deviation (Theft): Drivers stopping at unauthorized "Dhabas" or yards to offload 500kg of coal/ore.	Playback Forensic: Pick 5 random trips from the GPS portal. Replay the route history. Test: Look for "Stoppage Alerts" (> 10 mins) at non-designated locations. If a truck stopped for 30 mins in a remote area, it is a theft flag.
The "Ghost Trip" (Intra-Plant)	Fake Production: Internal trucks logging 20 trips/day between the Mine Face and Crusher to inflate production numbers, while actually making only 10 trips.	Cycle Time Analysis: Calculate the "Minimum Possible Cycle Time" (e.g., Load + Haul + Dump + Return = 30 mins). Test: If the log shows trips every 15 minutes (physically impossible), the operator is swiping cards or manually logging trips without moving.
FastTag Reconciliation	Fake Highway Trips: Transporter billing for a trip from Mine A to Port B (200km), but the truck never left the local area.	Toll Plaza Check: For long-haul logistics, demand the FastTag Statement. Test: If the invoice claims a trip on Date X, does the FastTag show a deduction at the specific Toll Plaza on that route? No Toll = No Trip.

Section 2: Transit Loss & Quantity Reconciliation

Focus: Did the material that left the source reach the destination?

Sub Process	Red Flag / Risk	Audit Procedure (Field Steps)
Source vs. Destination Weight	Pilferage: Material stolen en route. Source Weight = 20.00 MT, Destination Weight = 19.50 MT. The Transporter claims it is "within tolerance."	Shortage Recovery: Apply the formula: (Source- Destination) / Source %. Test: If the shortage > Tolerance (e.g., 0.5%), ensure the Total Value of Material (not just the excess loss) is recovered from the freight bill.
The "Watering" Fraud	Hiding Theft: Driver steals 500kg of Coal/Ore, then adds 500 liters of water before the destination weighbridge to match the weight.	Moisture Comparison: Compare the Source Moisture % vs. Destination Moisture % (from Lab Reports). Test: If Source Moisture is 8% and Destination is 12% (without rain), the driver added water. The weight of the water must be deducted from the billable weight.
Return Tare Check	Fuel Theft/Smuggling: Drivers filling the truck's hollow chassis with illegal scrap or fuel on the return (empty) leg.	Tare Discrepancy: Compare "Tare Weight at Destination (After Unloading)" vs. "Tare Weight at Source (Next Trip Entry)". Test: If the truck left the Port weighing 12T (Empty) but entered the Mine weighing 13T, they are smuggling something in.

Section 3: Freight Billing & Rate Management

Focus: Are we paying the correct price?

Sub Process	Red Flag / Risk	Audit Procedure (Field Steps)
Lead Distance Verification	Mileage Inflation: Paying for a "Lead Slab" of 40-45 km, when the actual shortest route is 38 km.	Google Maps vs. Odometer: Do not trust the Transporter's claim. Verify the distance using Google Maps or a physical Odometer check. Ensure payment is based on the "Shortest Motorable Route", not the route the driver chose to take.
Diesel Escalation (PVC)	Formula Manipulation: Transporter claiming price hikes when Diesel prices rise, but not passing on the benefit when Diesel prices fall.	Base Rate Check: Review the Freight PVC Calculation. Test: Check the "Base Diesel Price" used in the formula. Does it match the *actual* PSU pump price on the date of the invoice? Ensure "De-escalation" (reduction) is enforced when fuel prices drop.
Detention Charges (Demurrage)	Inefficiency Costs: Paying "Halting Charges" because trucks are waiting 24 hours at the mine gate to load.	Root Cause Audit: Analyze the "Time In" vs. "Time Out" logs. Test: If Average Turn Around Time (TAT) > 12 hours, investigate: Is the Loading machine broken? Is the Weighbridge slow? Detention charges are a penalty for *our* inefficiency.

Section 4: Vehicle Fitness & Compliance

Focus: Safety and Statutory Compliance.

Sub Process	Red Flag / Risk	Audit Procedure (Field Steps)
Tyre & License Swap	Safety Violation: Transporters using "bald tyres" or using one "Valid License" for multiple drivers.	Gate Surprise Check: Stop 5 trucks at the gate. 1. Check Tyre Tread Depth (must be > 2mm). 2. Check Driver's License (Does the photo match the face?). 3. Check Vehicle Insurance expiry.
Overloading Compliance	Legal Liability: Transporter overloading trucks to save fuel/trips. If caught by RTO, the Mine Owner (Consignor) is also liable.	Weighbridge Gross Weight: Check the Weighbridge log for "Gross Weight" > RLW (Registered Laden Weight). Test: If a 25T capacity truck is carrying 30T, flag it. The system should automatically block the gate pass for overloaded trucks.

17.4 Field Insight: The "Return Load" Opportunity

The Scenario: Trucks bring Raw Material (e.g., Coal) to the plant and return Empty.

The Opportunity: The same trucks can carry Finished Goods (e.g., Ash/Slag/Metal) on the return leg.

The Audit:

- Check if separate transporters are hired for Inward (Coal) and Outward (Ash).

- Recommendation: If the routes overlap, negotiate a "Back-Haul Rate" (usually 30-40% cheaper) by using the same fleet for both legs. Failure to do so is a "Synergy Loss."

17.5 Expert Insight: Key Frauds & Indicators

1. The "Water Tank" Tare Trick

- The Mechanism: As described in 17.2, manipulating the empty weight.
- Expert Indicator: A sudden spike in "Water Tanker" entries at the gate, or trucks dripping water *before* loading.
- Audit Fix: Install cameras at the weighbridge to check for "wet spots" under the truck during the Tare weighment.

2. The "GPS Simulator" (Intra-Plant)

- The Mechanism: Contractors are paid based on GPS distance traveled.
- The Fraud: Drivers use "GPS Spoofing" apps on their phones to simulate movement while the truck is parked, generating fake kilometers.
- Expert Indicator: GPS logs showing perfectly straight lines or uniform speeds (e.g., exactly 40kmph for 1 hour) which is physically impossible in a mine.

17.6 The Indian Scenario: Key Challenges

- "Speed Money": In Indian industrial belts, local goons often control the "Line" (queue) outside the plant. They create artificial congestion to force transporters to pay for entry.
- E-Way Bill Loophole: For Intra-plant movement (within the same premise/lease area), E-Way bills are often not generated. This creates a "Black Hole" where material can be moved out illegally without GST tracking.

Chapter 18: Outward Logistics (Rail/Rake & Road Dispatch)

This area is critical because "Logistics Cost" often eats up 20-30% of the sales margin. The primary risks here are Demurrage (Penalty), Dead Freight (Under-loading), and Quality Disputes (Grade Slippage)

The Value Chain: Sales Order > DO (Delivery Order) > Truck/Rake Placement > Loading > Weighment > Permit Generation (TP) > Dispatch

18.1 Introduction & Audit Objective

This is the "Cash In" chapter. Once the product leaves the gate, it becomes Revenue. The logistics usually involve Railways (Rakes) for bulk or Road Tippers for retail.

For the Layman: The goal is to maximize the load (sell more) without breaking the law (overloading) or paying penalties.

For the Expert Auditor: The goal is to optimize the "Payload" to avoid "Dead Freight" and ensure the "Grade" dispatched matches the "Grade" acknowledged by the customer.

Primary Audit Risks:

- Dead Freight: Paying freight for 4,000 Tonnes capacity but loading only 3,800 Tonnes.
- Punitive Freight: Overloading the wagons, attracting 6x penalty rates.
- Demurrage/Wharfage: Taking too long to load the train.
- Grade Slippage: Dispatching 64% Fe, but customer receiving 60% Fe (Theft en route).

18.2 Key Audit Focus Areas (Technical Concepts)

A. Railway Logistics: The "Carrying Capacity" Trap

Rail freight is charged on Carrying Capacity (CC) or Actual Weight, whichever is *higher*.

- Scenario A (Under-loading):
  - CC = 65 Tonnes. Actual = 60 Tonnes.
  - *Result:* You pay for 65T. You lost freight on 5T (Dead Freight).
- Scenario B (Over-loading):
  - CC = 65 Tonnes. Actual = 67 Tonnes.
  - *Result:* You pay normal freight for 65T + Punitive Freight (6x rate) for 2T.
- The Audit Sweet Spot: Loading exactly CC + Permissible Tolerance.

B. Demurrage & Wharfage

- Demurrage: Penalty for keeping the rake longer than the "Free Time" (e.g., 9 hours for loading).
- Wharfage: Penalty for not clearing the "Wharf" (platform) cargo within the stipulated time.
- Audit Check: Analyze the "Rake Detention Report." Is the delay due to "No Cargo" (Planning failure) or "Belt Breakdown" (Maintenance failure)?

C. Road Dispatch: Route Optimization

- The Risk: Transporters taking a longer route to bill higher freight, or taking a specialized route to bypass RTO checkposts (smuggling).
- The Control: Geofencing. The ERP should only auto-approve the bill if the GPS trace matches the "Approved Route" polygon.

18.3 Detailed Audit Work Program: Outward Logistics (Rail & Road)

Section 1: Rail Logistics (Rake Management)

Focus: Maximizing Wagon Capacity & Minimizing Penalties.

Sub Process	Red Flag / Risk	Audit Procedure (Field Steps)
Demurrage / Wharfage	Penalty Bleed: Taking > 9 hours to load a rake (Free Time). Indian Railways charges massive penalties per wagon per hour for delay.	Time-Log Analysis: Review the "Placement vs. Release" Report. Test: If average loading time is 12 hours (vs. 9 hours allowed), calculate the Total Demurrage Paid. Identify the root cause: Was the Silo broken? Was the Payloader slow? This is an avoidable loss.
Under-Loading (Dead Freight)	Capacity Loss: Paying freight for the "Carrying Capacity" (CC + 6 Tonnes) but loading 2 Tonnes less per wagon to be "safe." This is "Dead Freight" (paying for air).	Utilization Audit: Compare "Chargeable Weight" (what Railways billed) vs. "Actual Weight" (what we loaded). Test: If we consistently load 2T less per wagon, calculate the Opportunity Loss
Over-Loading (Punitive Charges)	Regulatory Fine: Loading > "CC+8". Railways impose Punitive Charges (6x to 10x of freight) for overloading.	Weighbridge Calibration: Check the "In-Motion Weighbridge" calibration logs. Is the automated "Cut-Off" system (which stops loading when the target weight is reached) working? Frequent punitive charges indicate the loading system is erratic.
Wagon Inspection (Rejection)	Operational Disruption: Railways placing "Damaged Wagons" (Bulged/Hole in floor)	Rejection Log: Check the "Wagon Rejection Register." Did the Loading Supervisor officially reject damaged wagons *before*

	which cannot be loaded, but counting them against the indent.	loading started? If not, Railways will charge freight for them even if they return empty.

Section 2: Road Dispatch (Sales)

Focus: Preventing Theft at the Weighbridge.

Sub Process	Red Flag / Risk	Audit Procedure (Field Steps)
Tare Weight Manipulation	Theft: Driver increases the empty truck's weight (Tare) by filling the water tank or hiding a heavy metal plate, then removes it before the Gross weight measurement.	"Double Weighment" Test: Conduct a surprise check. Re-weigh a truck *immediately* after it completes its "Tare" weight. Test: If the first Tare was 12,000 kg and your re-check shows 11,500 kg, the 500 kg difference is the intended theft amount.
Sequence Check (Missing Trips)	Revenue Leakage: Dispatching a truck without an invoice. The operator deletes the weighment record later.	Slip Sequence Audit: Check the "Weighment Slip Numbers" for continuity (e.g., WB-1001, WB-1002). Test: Are there missing numbers (e.g., WB-1003 is missing)? Ask for the "Cancelled Ticket Report." If a truck left with material on a "Cancelled Ticket," it is pure theft.
Turnaround Time (TAT)	Congestion Cost: Trucks waiting 10 hours inside the plant to load. This reduces the number of trips a	Gate-In vs. Gate-Out: Calculate the average TAT. Test: If TAT > 4 hours, identify the bottleneck (e.g., Security Check, Documentation, or Loading

	transporter can make, forcing them to hike rates.	Point). High TAT = Higher Freight Rates.

Section 3: Commercial & Quality Realization

Focus: Getting paid for what we sold.

Sub Process	Red Flag / Risk	Audit Procedure (Field Steps)
Grade Slippage (GCV Loss)	Revenue Loss: Billing the customer for "Grade G11" (4000 GCV) but the Third Party Sampler (TPS) declares it "Grade G13" (3400 GCV) at the destination. The Company has to issue a massive Credit Note.	Sampling Process Audit: Witness the "Joint Sampling" process. Test: Is the automatic sampler working? Are the samples sealed in Tamper-Proof Bags immediately? Compare "Internal Lab Result" vs. "TPS Result" for the last 50 rakes. A consistent negative variance implies our sampling is biased or the TPS is compromised.
Shortage Deduction	Customer Deduction: Customer deducting 2% for "Transit Shortage" when the contract allows only 0.5%.	Contract Reconciliation: Review the Sales Contract. Does it specify "Load Port Weight Final" or "Discharge Port Weight Final"? Test: If terms are "Load Port Weight Final," reject any shortage deduction by the customer.
E-Way Bill Compliance	Confiscation Risk: Dispatching goods with an E-Way Bill that expires before the	Expiry Log: Check the "E-Way Bill Extension" register.

	truck reaches the destination (due to breakdown/traffic).	Test: For trips > 500km, check if the validity was sufficient. If a truck was detained by GST authorities for an expired bill, check who paid the penalty (Transporter or Company?).

18.4 Expert Insight: Key Frauds & Indicators

1. The "Grade Swap" (The Road Mafia)

- The Mechanism: High-Grade Iron Ore (64% Fe) is dispatched. The truck stops at an unauthorized yard 50km away. The driver dumps the High-Grade ore and re-loads Low-Grade dust (55% Fe) mixed with water to match the weight.
- The Impact: The customer rejects the material or imposes a massive penalty. The "stolen" grade difference is pocketed by the mafia.
- Red Flag: Significant difference between "Mine Assay Report" and "Customer Receipt Assay Report," plus GPS halts >15 mins on the route.

2. The "Wet Coal" Dead Freight

- The Mechanism: Loading wet coal during monsoons.
- The Fraud/Loss:
    - *Loading:* 4,000T Wet Coal (Chargeable Wt: 4,000T).
    - *Journey:* Water evaporates.
    - *Destination:* 3,700T Dry Coal.
    - *Result:* Customer pays for 3,700T. Railway paid for 4,000T. Double Loss.
- Audit Fix: Install "Rapid Moisture Analyzers" on the loading belt to adjust the target loading weight dynamically.

18.5 Sales Contracts, Pricing & Counterparty Risk

The handbook is based in an environment where the mining process / operations are part of a larger organisation, and mine operates as captive plant. However, in case the mining operations is an individual profit centre or unit, key checks for customer contracts are mentioned hereunder.

The contract is the bridge between LB plate and customer bank account.

- Index-Linked Pricing & Formula Application
    - Audit Check: For long-term offtake contracts, verify that invoices correctly apply agreed indices (Platts/ODEX/IBM ASP), quality premia/discounts, freight terms and timing (lag months).
    - Red Flag: Manual spreadsheets "adjusting" index formulas with undocumented factors.
- Credit Risk & Dispatch Controls
    - Audit Check: Review credit limits, LC terms, security and the linkage between overdue status and dispatch release (system block when ageing exceeds threshold).
    - Risk: Sales continuing to over-leveraged or related-party customers with chronic overdues.
- Related-Party and Transfer Price Sales
    - Audit Check: Compare prices and terms for group entities vs independent customers of similar grade/volume.
    - Fraud Indicator: Systematic under-pricing to related trading entities that then capture margin downstream.

18.6 The Indian Scenario: Key Challenges

- Railway "Indenting" Corruption: Getting a rake in India often requires "Liasoning" with the Railway Board. Auditors often see cash expenses booked under "Incidental Expenses" or "Handling Charges" which are essentially bribes to get wagon allotment.

- The "Dalal" System: In road transport, the company rarely deals with truck owners. They deal with Brokers. The "Brokerage Commission" is often hidden inside the freight rate, inflating costs by 5-10%.
- Overloading Fines: Indian Supreme Court rulings are strict on overloading. A single overloaded truck can lead to the cancellation of the Transport Permit (TP). Auditors must ensure Zero Tolerance on overloading, even if it saves freight cost.

Part VI: Human Capital & Admin

In mining, manpower cost is the second largest expense after fuel/energy. It is also the area with the highest regulatory risk (Principal Employer Liability) and significant "Petty Cash" leakage in administration. In the mining and heavy industry sector, Payroll is not just about monthly salaries; it involves complex Production Linked Incentives (PLI), Piece-Rate Wages, and heavy Overtime (OT) components. This complexity makes it a breeding ground for financial leakage.

The Value Chain: Time & Attendance (Biometric) > Rostering > Incentive Calculation > Gross Pay Generation > Statutory Deductions (TDS/PF) > Bank Disbursement

19.1 Introduction & Audit Objective

Mining payroll is complex because it is rarely a fixed salary. It is heavily driven by Production Incentives, Overtime (OT), and Hardship Allowances.

For the Layman: The goal is to ensure people are paid what they are promised, but not more.

For the Expert Auditor: The goal is to audit the "Incentive Algorithm." If the Production Report is inflated to meet targets, the Payroll Department will unknowingly pay out millions in fake bonuses. The Audit must link Physical Production to Financial Payout.

Primary Audit Risks:

- Ghost Employees: Salaries paid to people who left months ago or never existed.
- Incentive Fraud: Manipulating production data to trigger higher bonus slabs.
- Overtime (OT) Abuse: Supervisors rostering staff for OT when there was no work, often for a kickback ("Cut").

19.2 Key Audit Focus Areas (Technical Concepts)

A. The Production Incentive Audit

- The Concept: Most mines pay a variable bonus based on achieving monthly targets (e.g., Excavation > 1 Million BCM).
- The Leakage:
  - *Scenario:* Target = 100. Actual = 95.
  - *The Fraud:* The Survey team "adjusts" the closing stock measurement to show 101.
  - *The Result:* The entire workforce gets a 20% bonus. The company pays cash for "Paper Production."
- Audit Check: Compare the Production Bonus Payout vs. Sales/Dispatch. If Production is consistently meeting targets but Sales are lagging and Stockpiles aren't growing, the production figures are fake.

B. Overtime (OT) Analysis

- The Benchmark: In a 3-shift operation, OT should be < 5% (only for emergency absenteeism).
- Red Flag: If specific departments (e.g., Workshop) consistently show 30% OT, it indicates either under-staffing or systematic fraud where employees work slow during the shift to create OT needs.

19.3 Detailed Audit Work Program: Payroll & Incentives

Section 1: Time, Attendance & Overtime (The Input)

Focus: Paying for time actually worked.

Sub Process	Red Flag / Risk	Audit Procedure (Field Steps)
Biometric vs. Manual	The "Regularization" Loophole: High frequency of "Missed Punch" approvals.	System Log Audit: Extract the "Manual Regularization Report" from the HRMS.

	Employees deliberately avoid punching in/out and get supervisors to "regularize" their attendance to hide late arrivals or early exits.	Test: If an employee has >3 manual adjustments per month, investigate their actual physical presence (CCTV, PC Login logs, or Gate Access logs).
Overtime (OT) Fraud	The "Lifestyle" OT: Drivers, Peons, or Operators consistently booking 4 hours of OT every single day, effectively treating it as a fixed salary component.	Gate Log Correlation: Select the "Top 10 OT Earners." Compare their claimed OT hours with the Biometric "Out" Time. Test: If the biometric shows they left at 6:00 PM but the OT sheet claims work until 9:00 PM, it is fraud.
Underground Allowance	Location Fraud: Surface staff claiming "Underground Allowance" (which is higher) without actually going below ground.	Cap Lamp Register: In underground mines, every person going down must take a "Cap Lamp." Check the Lamp Room Register. If the employee claimed the allowance but no lamp was issued to them on those dates, the claim is fake.
Negative Leave Balance	Recovery Failure: Employees taking more leave than accrued, resulting in a negative balance that is never recovered.	Leave Ledger Scrutiny: Check for employees with "Negative Leave" balances carried forward for >6 months. Ensure "Loss of Pay" (LOP) was deducted for the excess days.

Section 2: Production Incentives & Variable Pay

Focus: Linking pay to genuine output.

Sub Process	Red Flag / Risk	Audit Procedure (Field Steps)
Incentive Calculation	Data Manipulation: Production Managers rating their own team as "Outstanding" (100% Payout) even when the mine missed its monthly target.	Bell Curve & KPI Check: Compare individual ratings with the overall Mine Performance Score. Test: If the Mine's production was 80% of the target, the "Production Bonus" should mathematically be lower. If everyone received 100%, the formula was overridden.
The "Availability" Game	Metric Inflation: Maintenance teams are paid bonuses based on "Equipment Availability %." They delay opening "Job Cards" for broken machines to make them appear available longer.	Downtime Reconciliation: Compare the "Production Log" (Operator's notes on breakdown) vs. the "Engineering Log" (Job Card opening time). Test: If the Operator noted "Breakdown at 10:00 AM" but the Job Card was opened at "2:00 PM," the 4-hour gap is fraud to boost the Availability Bonus.
Piece-Rate Wages	Ghost Output: In manual mines, workers are paid per tub/tonne of coal cut. "Ghost Tubs" are recorded to inflate wages.	Tub Count vs. Survey: Compare the "Total Tubs Paid" (Payroll data) vs. the "Surveyed Volume" (Engineering data) for the month. Test: If Payroll paid for 5,000 tubs but the Survey shows only 4,500 tubs' worth of excavation, the difference is phantom payments.

Section 3: Payroll Processing & Controls

Focus: Preventing payments to ghosts.

Sub Process	Red Flag / Risk	Audit Procedure (Field Steps)
Ghost Employee Check	Salary Theft: HR Manager creating a fake employee record and routing the salary to their own (or a relative's) bank account.	Bank Account Duplicate Test: Run a duplicate check on Bank Account Numbers in the Payroll Master. Test: If two different Employee IDs share the same Bank Account Number, it is a confirmed fraud.
The "Leaver" Loophole	Post-Exit Payment: An employee resigns on the 15th, but Payroll processes the full month's salary because the "Exit Date" wasn't updated in SAP.	Resignation vs. Bank Transfer: Compare the "List of Resigned Employees" (Exit Interview log) with the "Bank Transfer Sheet" for the subsequent month. Ensure no salary was credited after the Last Working Day.
Salary Fitment	The "Pay Hike" Anomaly: A new hire offered a 60% hike on their last drawn salary without justification, violating the 30% cap policy.	Peer Comparison: Compare the new hire's CTC with the "Internal Peer Group" (employees in the same grade). If the new hire is paid significantly more without niche skills, flag it as potential nepotism or kickbacks.

Section 4: Statutory Compliance (Mining Context)

Focus: Legal liability.

Sub Process	Red Flag / Risk	Audit Procedure (Field Steps)
Form B Register	Legal Exposure: Employees working on-site but not listed in the	Spot Check: Randomly pick 5 workers on the mine face. Ask for their names. Check if they are

	statutory "Form B" (Register of Employees). If an accident occurs, the company is in severe legal trouble.	listed in the Form B Register kept at the Mine Manager's office.
CMPF / PF Deductions	Liability Risk: Deducting Provident Fund (Coal Mines PF or EPF) but delaying the deposit to the government.	Challan Date Check: Verify the Date of Deposit on the PF Challan. Any delay attracts interest and penalties which are disallowed expenses.
Gratuity Valuation	Under-Provisioning: Booking Gratuity liability based on a simple Excel calculation instead of an Actuarial Valuation.	Actuarial Report Match: Verify if the provision in the Balance Sheet matches the "Actuarial Valuation Report." Ensure the data sent to the actuary (DOB, Basic Salary) was accurate.

19.4 Field Insight: The "Mustor Roll" vs. "Token" System

The Scenario: In underground mines, attendance is a life-and-death matter.

- The Control: The "Token System." Every miner picks up a brass token when entering the cage/lift and returns it when coming up.
- The Audit:
    - Go to the Lamp Room / Time Office at the end of the shift.
    - The Check: If the "Attendance Register" (Mustor Roll) marks an employee as "Present" (P), but their Token is still hanging on the board (meaning they didn't go down), it is a fake attendance marking.

- Implication: They are claiming wages (and underground allowance) while sitting at home or working elsewhere.

19.5 Expert Insight: The "Full & Final" Trap

- The Fraud: An employee resigns. The "Full & Final Settlement" (FnF) is processed.
- The Glitch: The employee is removed from the "Active List" but not the "Bank Advice List." The company continues to pay his monthly salary for 3 months, which is shared between the ex-employee and the Payroll Executive.
- Audit Fix: Reconcile the "Exit Interview Log" with the "Bank Transfer List" monthly.
- (Similarly check reimbursements for exited employees)

This chapter differs from Chapter 15 (Procurement). Chapter 15 was about "Billing." Chapter 20 is about "Statutory Compliance" and "Human Rights." he primary risks here are "Ghost Workers" (billing for men who aren't there), Wage Theft (contractors underpaying workers), and Legal Liability (Sham Contracts).

The Value Chain: Labor License > Gate Pass Issuance > Daily Attendance > Wage Sheet Verification > PF/ESI Deposit > Invoice Certification

20.1 Introduction & Audit Objective

For the Layman: Ensuring the contractor pays the workers their legal dues (PF, ESI) so the government doesn't penalize the mine owner.

For the Expert Auditor: The objective is to mitigate "Principal Employer Liability." Under Indian Law (CLRA Act), if the contractor defaults, the Mine Owner pays.

Primary Audit Risks:

- PF/ESI Theft: Contractor deducts 12% from worker but doesn't deposit it with the Govt.
- Minimum Wage Violation: Paying below statutory rates in cash.
- Double Employment: The same worker shown in two different contracts to inflate man-count.

20.2 Key Audit Focus Areas (Technical Concepts)

A. The TRRN Validation (The Gold Standard)

Do not accept photocopies of PF Challans. They are easily photoshopped.

- The Expert Check:
    1. Get the TRRN (Temporary Return Reference Number) from the Challan.
    2. Go to the EPFO Website (Public Search).

3. Verify if the amount was *actually credited* to the bank.
4. Download the "ECR" (Electronic Challan Return) to see the *list of names*.
   - *Fraud:* Contractor pays PF for 10 people but bills the company for 100 people.

B. The "Split Wage" Structure

- The Trick: Contractors keep the "Basic Wage" low (e.g., ₹5,000) and "Allowances" high (e.g., ₹10,000) to reduce the 12% PF liability.
- The Law: Recent Supreme Court rulings (Surya Roshni case) state that allowances which are "universally paid" must be part of Basic Wages for PF calculation.
- Audit Check: Ensure PF is deducted on Basic + DA + Retaining Allowance.

20.3 Detailed Audit Work Program: Contract Labor Management

Section 1: Onboarding & Identity Management

Focus: Are the workers real and legally allowed to work?

Sub Process	Red Flag / Risk	Audit Procedure (Field Steps)
Identity Verification	Proxy Workers: The person working on-site is not the person listed in the gate pass/records. This happens when a verified worker sends a younger/unverified relative as a "proxy."	The "Face-to-Photo" Check: Conduct a surprise muster parade at the worksite. Test: Randomly pick 10 workers. Check their Gate Pass Photo. Does it match the face? Ask them their name and father's name. Mismatches indicate "Proxy Labor."
Statutory Licensing	Capacity Breach: The contractor holds a Labor License (Form VI) for 50 workers but has deployed 80. The	Headcount vs. License: Compare the "Daily Gate Entry Report" (Total Contract Workers) vs. the

	Principal Employer is liable for prosecution for the excess 30.	"Labor License Limit" of each contractor. Test: If the count exceeds the license limit, stop the bill payment immediately.
Medical Fitness (Form O)	Health Liability: Hiring workers unfit for heavy mining tasks (e.g., poor vision, hernia). If an accident occurs, the mine is liable.	Form O Audit: Check the "Form O" (Medical Certificate) for sample of new joiners. Test: Was the check-up done by the Company Doctor or an external clinic? Verify the date of exam vs. date of joining. PME (Periodical Medical Exam).

Section 2: Attendance, Timekeeping & Ghost Workers

Focus: Preventing "Air Billing" (billing for non-existent work).

Sub Process	Red Flag / Risk	Audit Procedure (Field Steps)
Biometric vs. Manual	Ghost Workers: The contractor submits a manual "Muster Roll" with 100 names. The Biometric system shows only 80 punches. The 20 difference is fake billing.	System Reconciliation: Never pay based on manual sheets. Test: Compare the Invoice Headcount vs. Biometric "In" Punch Report. Disallow any name that lacks a corresponding electronic gate entry.
The "Double Shift" Fraud	Duplicate Billing: A worker punches "Out" from Shift A and immediately punches "In" for Shift B under a different contractor name/ID.	Duplicate Logic Check: Run a script on the Time Office data: "Same Name + Same Father's Name" or "Same Biometric Template" active in two consecutive shifts. Physically check if the person is capable of working 16 hours straight.
Proxy Punching	Attendance Manipulation: The	CCTV / Gap Analysis: Review the "Punch Logs."

	Supervisor carries 5 ID cards and punches them all one by one at the turnstile.	Test: Look for "Rapid Fire Punches" (e.g., 5 punches within 10 seconds at the same terminal). This is physically impossible for 5 different people. It proves one person is swiping multiple cards.

Section 3: Wages & Statutory Compliance (The "ATM Scam")

Focus: Ensuring the worker actually gets the money.

Sub Process	Red Flag / Risk	Audit Procedure (Field Steps)
The "ATM Card" Scam	Wage Theft: The contractor opens bank accounts for workers but keeps the ATM cards/Passbooks himself. He withdraws the full wage, pays the worker a cash pittance, and pockets the rest.	Worker Interview: Randomly interview 5 workers away from their supervisor. Test: Ask: "Do you have your ATM card with you?" "Do you know your PIN?" "How much was credited last month?" If they don't have the card or don't know the amount, the contractor is stealing wages.
Minimum Wages & VDA	Underpayment: Paying "Unskilled" rates to "Skilled" operators (e.g., Dumper Drivers) to save costs.	Skill Mapping: Pick 5 Dumper/Excavator operators. Check their designation in the Wage Register. Test: If they are listed as "Helper" or "Unskilled," it is a violation of the Minimum Wages Act. They must be paid "Skilled" or "Highly Skilled" rates.
PF & ESI Liability	Theft of Contribution: Contractor deducting 12% PF from wages but not depositing it.	ECR Verification: Do not accept the "Challan Copy" (easily forged). Demand the Electronic Challan Return (ECR) which lists worker names.

		Test: Check if the specific names from the invoice are present in the ECR list.

Section 4: The "Sham Contract" Risk

Focus: Avoiding the "Regularization" Trap.

Sub Process	Red Flag / Risk	Audit Procedure (Field Steps)
Direct Supervision	Legal Risk: Company managers directly supervising contractor workers, approving their leave, and assigning tasks. This proves a "Master-Servant" relationship, allowing workers to claim permanent employment.	Instruction Chain Check: Observe the shop floor. Test: Who gives the daily briefing? If a Company Officer gives the briefing directly to workers (bypassing the Contractor Supervisor), it is a "Sham Contract" indicator. Leave forms should only be signed by the Contractor.
Tool Issue	Asset Ownership: Company issuing personal tools (spanners, drills) directly to contractor workers.	Tool Store Log: Check the "Tool Issue Register." Test: Tools should be issued to the Contractor Firm, not individual workers. If workers hold company gate passes for tools, it indicates they are treated as employees.
Perennial Nature	Statutory Violation: Using contract labor for "Core" permanent activities (prohibited under CLRA Section 10).	Role Audit: Are contract workers doing the exact same job as permanent workers on the same assembly line? If yes, it violates the CLRA Act prohibition on contract labor in core perennial activities.

Section 5: Welfare & Separation

Focus: Human rights and final settlements.

Sub Process	Red Flag / Risk	Audit Procedure (Field Steps)
Full & Final (F&F)	Retention Money: Contractor failing to pay the "Retrenchment Compensation" or "Gratuity" (if >5 years) when a contract ends.	Exit Interview: Interview 2 workers who left recently. Test: Did they receive their final settlement? Contractors often pocket the unpaid wages of workers who leave abruptly.
Canteen & Creche	Facility Denial: The law mandates a Canteen (if >100 workers) and Creche (if >50 women). Company charges contractor workers higher rates for food than permanent staff.	Subsidized Food Audit: Check the Canteen Rates. Test: Discrimination in food rates or quality between permanent and contract staff is illegal and a major industrial relations risk.

Chapter 21: Administration, Camp & Security Management

Value Chain: Facility Mgmt > Vendor Selection > Service Delivery (Food/Transport) > Consumption Logging > Invoice Verification

21.1 Introduction & Audit Objective

Mines are often in remote jungles. They operate "Townships" or "Camps" for staff. This involves running Canteens, Guest Houses, Fleets of cars, and Security details.

Primary Audit Risks:

- Canteen Pilferage: Booking "Ghost Meals" to inflate the subsidy bill.
- Camp Diesel Theft: Generators running on paper when grid power is available.
- Security Collusion: The guards are often part of the theft ring.

21.2 Key Audit Focus Areas

A. Canteen Management (The Subsidized Meal)

- The Model: Meal Cost = ₹50. Employee pays ₹5. Company pays ₹45.
- The Fraud: The caterer claims 1,000 plates were served. Actual consumption was 600. The company pays subsidy on 400 ghost meals (400 x ₹45 = ₹18,000 loss per day).
- The Control: "Coupon or Biometric" based food issue. No manual headcounts.

B. Security Management (Guarding the Guards)

- The Risk: Security agencies billing for "10 Guards per shift" but deploying only 6. The other 4 are "sleeping partners" (names on roster, salary shared).
- The Audit: Surprise Night Checks. Count the physical posts.

- The Scrap Connection: Check the Security Gate Logbook for "Returnable Items." If items go out but don't come back, and Security doesn't flag it, they are complicit.

21.3 Detailed Audit Work Program: Admin, Camp & Security

Section 1: Camp & Colony Management (The "Township" Risk)

Focus: Managing residential assets and utility costs.

Sub Process	Red Flag / Risk	Audit Procedure (Field Steps)
Quarter Allotment	Illegal Subletting: Employees allotted company quarters who secret rent them out to outsiders for cash, or retain the house long after transfer/resignation.	Door-to-Door Census: Pick 10 random quarters from the "Allotment Register." Physically visit them in the evening. Test: Who opens the door? If the occupant is not the employee listed in the register, it is a subletting fraud.
Electricity Recovery	Revenue Leakage: Employees consuming unlimited power (ACs/Heaters) because the "Per Unit Recovery" is not deducted from their salary.	Meter vs. Payroll Check: Select 10 high-consumption quarters. Check the Meter Reading for the last month. Test: Check the Salary Slip of the occupant. Was the electricity charge deducted? If the meter is "Faulty" for >3 months, it is often intentional.
Diesel Generator (DG) Fuel	Diesel Theft: Remote camps run on DG sets. Admin claims 20 Liters/Hour consumption, but	SFC Validation: Calculate Specific Fuel Consumption (SFC).

	actual load requires only 12 Liters/Hour. The difference is siphoned.	

Section 2: Guest House & Travel Management

Focus: Controlling "Lifestyle" expenses.

Sub Process	Red Flag / Risk	Audit Procedure (Field Steps)
Guest House Inventory	Consumable Theft: High consumption of toiletries, towels, and food. Staff stealing new inventory and claiming it was "used" by guests.	Stock vs. Occupancy Ratio: Calculate consumption per room night. Test: If the register shows 10 room nights but 50 soap bars/towels were issued, the inventory is leaking. Physically count the "Linen Stock" against the Master Register.
External Hotel Booking	Entitlement Breach: Junior employees staying in 5-Star hotels or booking "Suites" in violation of the Travel Policy limits.	Bill Audit: Select 10 high-value travel reimbursements. Test: Compare the Room Rate Charged vs. the Employee Grade Limit. If a Supervisor (Limit ₹3,000) stayed in a room costing ₹7,000, ask for the "Special Approval" note.
Corporate Rate Usage	Financial Loss: Employees booking hotels directly online (e.g., Booking.com) at market rates, bypassing the	Rate Comparison: Check the invoice.

	company's negotiated "Corporate Rates."	
"No-Show" Charges	Wasteful Expenditure: Booking rooms/flights but failing to cancel them in time, resulting in 100% cancellation charges.	Ledger Scrutiny: Filter the General Ledger for "Cancellation Charges." Test: Identify repeat offenders. Was the business reason for cancellation valid (e.g., meeting cancelled), or was it pure negligence?

Section 3: Field Clinic (Occupational Health Center)

Focus: Employee safety and drug inventory control.

Sub Process	Red Flag / Risk	Audit Procedure (Field Steps)
Medicine Expiry Control	Safety Hazard: Dispensing expired medicines to workers. This is a severe negligence risk.	Shelf Inspection: Visit the Clinic Store. Check the Expiry Dates on randomly selected strips (Antibiotics, Painkillers). Test: Are expired medicines kept in a separate "Red Bin"? If they are mixed with fresh stock on the shelf, it is a high-risk observation.
Bio-Medical Waste	Statutory Violation: Dumping used syringes and bloody cotton in the general dustbin. This attracts huge penalties from the Pollution Control Board.	Disposal Trail Check: Check the "Bio-Medical Waste Register." Test: Does the quantity generated match the "Collection Receipt" from the authorized disposal agency? Verify if color-coded bins (Yellow/Red/Blue) are used correctly on site.

Ambulance Availability	Emergency Failure: The ambulance is used for admin work (ferrying veggies/staff) and is not available during a mine accident.	Logbook Review: Check the Ambulance Logbook. Test: Are there trips to the "Market" or "Railway Station"? An ambulance should strictly be used for patients only. Check if the Oxygen Cylinder inside is full.
Fake Medical Certificates	Absenteeism Fraud: The Company Doctor issuing "Sick Leave" certificates to employees who just want a holiday, in exchange for favors.	Pattern Analysis: List employees with >3 Sick Leaves per month. Test: Cross-verify with the "OPD Register." Did they actually visit the clinic on the day they claimed sick leave? If there is no OPD entry but a certificate exists, it is fake.

Section 4: Security Operations

Focus: Is the Guard actually Guarding?

Sub Process	Red Flag / Risk	Audit Procedure (Field Steps)
Deployment vs. Billing	Ghost Guards: The Security Agency bills for 50 guards/shift. In reality, only 35 are present. The Agency pockets the wages of the missing 15.	Night Parade: Conduct a Surprise Check at 2:00 AM. Assemble all guards (except critical posts). Test: Count the heads. Compare with the "Shift Duty Roster." If Roster says 50 but you find 35, disallow the difference from the invoice.
Scrap Movement Control	Collusion: Security guards allowing scrap trucks to leave without	CCTV Playback: Pick a random "Gate Pass" for scrap. Watch the CCTV footage of that specific time.

	checking papers, in exchange for a bribe.	Test: Did the guard physically climb the truck to inspect? Did they check the driver's cabin? If the truck just rolled through, the guard is compromised.
Key Management	Theft Risk: Keys to critical areas (Stores, Server Room) kept hanging openly in the Security Booth, accessible to anyone.	Key Register Audit: Check the "Key Issue Register." Test: Are keys returned every shift? Are unauthorized persons signing out keys? Verify if "Duplicate Keys" are sealed in an envelope or lying loose.

21.4: Field Insight- The "Mock Drill"

The Scenario: You suspect Security is lax at the main gate.

The Test: The "Dummy Intrusion" (Authorized by GM).

1. The Setup: Instruct a known driver to try and take a "Dummy Laptop" (without a gate pass) out of the gate in the boot of a car.
2. The Observation: Stand at a distance and watch.
3. The Result:
    - Pass: Guard checks the boot, finds the laptop, stops the car.
    - Fail: Guard waves the car through without checking.
4. Audit Finding: "Ineffective Access Control." This provides undeniable evidence of security failure.

21.5 Expert Insight: The "Vegetable" Smuggling

- The Fraud: In a remote mine, the "Vegetable Supply Truck" for the canteen entered every morning.
- The Discovery: The truck had a false bottom. It brought in 50kg of vegetables but smuggled out 200kg of Copper Scrap/Cable cuttings every day. Security never checked the "Empty" truck going out because it was a "kitchen vendor."
- Audit Lesson: All vehicles, including ambulances, VIP cars, and food trucks, must be subject to random checks.

Part VII: Finance & Treasury

Chapter 22: Accounting, Taxation & Books Closure

This chapter serves as the "Scorecard" audit—verifying that the operational realities discovered in previous chapters (Production, Logistics, Procurement) are correctly translated into Financial Statements.

The Value Chain: Transaction Capture (ERP) > Sub-Ledger Reconciliation > Accruals & Provisions > Taxation Logic > Financial Reporting

22.1. Process Overview & Objectives

The "Record to Report" (R2R) process in mining is unique due to the concept of "Wasting Assets" (Depletion) and the heavy burden of statutory levies (Royalty/DMF). The objective is not just to tally the Debit/Credit, but to ensure the Balance Sheet reflects the true remaining life of the mine and its liabilities.

The Value Chain:

Transaction Capture (ERP) > Sub-Ledger Reconciliation > Accruals & Provisions > Taxation Logic > Financial Reporting

Process KPIs (Management Dashboard):

- Days to Close: Time taken to freeze books (Benchmark: Work Day + 4).
- Post-Closing Adjustments: Number of entries passed *after* the draft Trial Balance (High count = Poor process).
- Effective Tax Rate (ETR): Total Tax / Profit Before Tax.
- Provision Utilization: Actual payout vs. Provision created (Variance >10% indicates poor estimation).

22.2. The "Why" Box: Auditor's Mindset

Auditor's Mental Model:

"In Operations, we looked for Leakage (Theft). In Accounting, we look for misrepresentation (Lies).

Mining managers are under immense pressure to show profit and reduce 'Cost per Tonne'.

- If Profit is Low: They will try to Capitalize expenses (treat repairs as assets) or over-value the closing stock (treat waste as valuable ore).
- If Profit is High: They might over-provision (create secret reserves) or accelerate depreciation to save tax.

The Golden Rule: The Accounts must mirror the Physical reality found in Chapters 5-18. If the Mine Plan says the pit is getting deeper (more waste), the Cost of Production *must* go up. If it doesn't, the accounts are cooked."

22.3. Common Industry Jargon

- Stripping Ratio (SR): The ratio of Waste removed to Ore recovered. Used to allocate costs.
- IFRIC 20: The global accounting standard determining when to capitalize Overburden Removal (Stripping Activity Asset).
- SRF (Site Restoration Fund): A mandatory escrow account for future mine closure expenses.
- RCM (Reverse Charge Mechanism): GST rule where the *recipient* pays the tax (Applicable on Royalty & DMF).
- NRV (Net Realizable Value): Estimated selling price minus cost to sell.
- CWIP (Capital Work in Progress): Assets under construction.

22.4. Key Regulatory & Technical References

- Accounting Standards (Ind AS / IFRS):
    - Ind AS 16 / IAS 16: Property, Plant & Equipment (Component Accounting).
    - Ind AS 2 / IAS 2: Inventory Valuation.

    - Ind AS 37: Provisions, Contingent Liabilities (Mine Closure).
    - IFRIC 20: Stripping Costs in Production Phase.
- Statutes (Indian Context):
    - Section 43B (Income Tax Act): Statutory dues (Royalty/PF) allowed as expense only on *payment*, not accrual.
    - MMDR Act: Calculation of Royalty, DMF (District Mineral Foundation), and NMET.
    - GST Act: Section 17(5) (Blocked Credits) and RCM provisions.

22.5 Detailed Audit Work Program: Accounting, Taxation & Books Closure

This program serves as the "Scorecard Audit," verifying that the operational realities discovered in previous chapters (Production, Logistics) are correctly translated into Financial Statements. The primary risks involve "Wasting Assets" accounting, Stripping Ratio manipulation (IFRIC 20), and Statutory Tax Leakages.

Section 1: Provisions, Accruals & Liability Management

Focus: Completeness of Liability (Are all expenses booked?)

Sub Process	Red Flag / Risk	Audit Procedure (Field Steps)
Cut-off Procedures (GRN vs. Booking)	Profit Inflation: Material received on March 30th but booked in April to shift the expense to the next year and inflate current year profit.	Inward Register Trace: Review the "Security Inward Register" for the last week of March. Trace these entries to the Goods Receipt Note (GRN) and the Ledger. Test: If a truck entered on Mar 30th but the liability was booked in April, it is a cut-off error.

Power & Fuel Accrual	Under-Accrual: Booking power cost based on an "Average" estimate instead of the actual meter reading, leading to variance.	Meter Reading Reco: Verify the Physical Meter Reading taken on March 31st. Reconcile this with the "Provision for Power" entry. Test: Does Provision = (Units consumed X Rate)? If provision is just a lump sum "average," it is incorrect[6].
Contractor Service Liability	Hidden Liabilities: Work completed in March (e.g., Transport, Excavation) but not booked because the engineer signed the "Work Completion Certificate" (WCC) in April.	WCC & Logbook Check: Select major service contracts. Check the Work Completion Certificates for the last week of March. Test: Was a provision created for work done but not yet certified?.
The "Phantom" Provision	Income Smoothing: Creating a lump-sum "General Provision" (e.g., $1M) in good months and reversing it in bad months to artificially smooth out profit volatility.	Basis of Provision Test: Scrutinize all "General" or "Ad-hoc" provisions. Check the detailed calculation sheet. Test: If the provision is a round number with no backing working paper, it is a "Cookie Jar Reserve".

Section 2: Inventory Valuation (The "Value" Audit)

Focus: Is the stock valued at the lower of Cost or Net Realizable Value (NRV)?

Sub Process	Red Flag / Risk	Audit Procedure (Field Steps)
Low-Grade Ore Valuation	Asset Overstatement: Valuing low-grade ore (e.g., <50% Fe, which is effectively waste) at the "Full Cost of Production," inflating asset value.	NRV Test: Compare the Book Cost per Tonne vs. Recent Market Sales Price for *each specific grade*. Test: If Cost > Sale Price (or if the grade is effectively waste), the stock must be written down to NRV.
Overhead Allocation	Cost Inflation: Including "Admin Salaries," "Selling Expenses," or "CSR" in the Inventory Cost to inflate profit (Capitalizing expenses into stock).	Cost Sheet Review: Review the "Cost of Production" (COP) sheet used for valuation. Test: Ensure only Production Overheads are included. Exclude Admin/Selling overheads strictly as per Ind AS 2.
Non-Moving Stock Provision	Dead Stock: Carrying rusted or weathered coal/ore at full value in the books.	Ageing Analysis: Filter inventory for items aged > 1 year. Test: Has a provision for obsolescence been created? Physical inspection usually reveals degradation requiring a write-down.
NRV Basis Verification (The "Phantom Bidder" Test)	Valuation Fraud: Valuing a stockpile of waste/low-grade rock at high rates based on a single "Letter of Intent" or "Quote" from a contractor who never actually bought the	The "Subsequent Sales" Trace: 1. Identify any inventory valued based on "Quotes" rather than active price lists. 2. Check the Sales Register for the 6 months *following* the valuation date.

	material. This inflates inventory value with no real realizable cash flow. (Ind AS 2 / IAS 2 Principle: NRV is the "Estimated selling price in the ordinary course of business." A single, unexecuted quote does not constitute "ordinary course.")	3. Test: Did that specific bidder actually purchase and pay for at least 10% of the stock at that quoted price? Result: If Zero Sales materialized, the quote is invalid evidence of NRV. The stock must be valued at Zero or purely nominal scrap value.

Section 3: Fixed Assets & Depreciation (The "Capital" Audit)

Focus: Correct Capitalization and Amortization.

Sub Process	Red Flag / Risk	Audit Procedure (Field Steps)
Repairs vs. Capitalization	Capex Fraud: Booking routine repairs (e.g., "Engine Overhaul," "Servicing") as "Capital Additions" to reduce Opex and boost P&L.	Voucher Scrutiny: Scan the "Plant & Machinery Additions" ledger. Look for keywords like "Repair," "Service," "Overhaul." Test: Capitalize *only* if the expense increases capacity or extends useful life beyond the original standard.
Component Accounting	Depreciation Error: Depreciating a heavy dumper (15-year life) as a single asset, ignoring that its Engine (5 years) and Tyres (1 year) wear out faster.	Depreciation Schedule Check: Check the Fixed Asset Register for heavy machinery. Test: Are major components (Engine, Chassis, Tyres) listed separately with different useful lives? If not, depreciation is understated.

CWIP Parking	Hidden Impairment: Failed or stalled projects (e.g., a stopped crusher project) kept in CWIP for >3 years to avoid booking the loss/impairment in P&L.	CWIP Aging Analysis: Filter CWIP for items > 2-3 years old. Interview Project Managers. Test: Is work actually progressing? If the site is abandoned, the CWIP must be impaired/written off immediately.
Mining Rights Amortization	Amortization Lag: Amortizing "Leasehold Land/Rights" on a Straight Line (Time) basis instead of "Unit of Production" (Depletion).	Depletion Check: Check the amortization method. Test: It must be based on Tonnes Extracted / Total Reserves. If you mine 50% of reserves in Year 1, you must expense 50% of the land cost, not 1/20th.

Section 4: Taxation & Statutory Compliance

Focus: Preventing Penalties and Interest.

Sub Process	Red Flag / Risk	Audit Procedure (Field Steps)
GST on Royalty (RCM)	Litigation Risk: Failing to pay GST on Royalty and District Mineral Foundation (DMF) payments under Reverse Charge Mechanism (RCM).	RCM Liability Check: Verify the RCM Liability Register. Test: Does the taxable value match the Royalty/DMF paid? This is a common litigation point; ensure liability is booked and paid.
ITC Reversal (Sec 17(5))	Tax Leakage: Failing to reverse Input Tax Credit (ITC) for goods lost, stolen, or written off.	Write-Off Ledger Check: Review the "Write-off" / "Theft" GL codes. Test: Check if the corresponding GST Input Tax Credit was

		reversed as required by Section 17(5).
TDS on Transport	Compliance Error: Deducting 1% TDS (Individual) instead of 2% (Company) due to misclassification of the transporter's status.	Vendor Master Check: Verify the "Deductee Type" in the Vendor Master against the transporter's PAN status. Test: Ensure the correct TDS rate is applied to payments.
Transfer Pricing (RPT)	Profit Shifting: Selling ore to a sister company (Related Party) at a price significantly lower than the market rate to shift profit to a tax-exempt unit.	Arm's Length Test: Compare the "Transfer Price" of ore sold to the group entity vs. the "Market Price" sold to unrelated third parties in the same month. Document the variance.

Section 5: Management Reporting (MIS) & Cut-Off

Focus: No "Window Dressing" of the financial results.

Sub Process	Red Flag / Risk	Audit Procedure (Field Steps)
Sales Cut-Off (The "April" Sale)	Revenue Inflation: Booking sales on March 31st (11:59 PM) to meet targets, for trucks that actually left the gate in April.	Gate Log Trace: Pick the last invoices of March 31st. Test: Match Invoice Date vs. E-Way Bill Date vs. Weighbridge Stamp. If the truck left after midnight, the revenue belongs to April. Reverse it.
Stripping Ratio Fraud Indicator	Profit Manipulation: Accounts show a stripping ratio of 1:2 while the Mine Plan says 1:4. The cost of the missing 2 tons of waste	Ratio Comparison: Compare the "Stripping Ratio" in the Cost Sheet vs. the "Geological Stripping Ratio" (Plan).

	has been fraudulently capitalized.	Test: A significantly lower SR in accounts indicates aggressive capitalization of Opex.

22.6. Common Audit Observations

1. The "Phantom" Provision: Management creates a lump-sum "General Provision" during good months and reverses it during bad months to smooth out profits. (Income Smoothing).
2. CWIP Parking: Failed projects (e.g., a stalled crusher installation) kept in Capital Work In Progress (CWIP) for 3+ years to avoid charging depreciation or impairment loss.
3. Royalty Calculation Error: Paying Royalty on the *Invoiced Price* instead of the *IBM/Published Price* (when IBM price is higher), leading to short payment and penalties.

22.7. Key Frauds & Indicators

Fraud 1: The Stripping Ratio Manipulation (IFRIC 20 Abuse)

- The Concept: Removal of waste (Overburden) is a cost. However, if it exposes *future* ore, it can be Capitalized.
- The Fraud: The company treats *routine* waste removal as *development* waste removal.
- The Impact: Millions of dollars of Opex are moved to the Balance Sheet (Asset). Profit skyrockets.
- Indicator: The "Stripping Ratio" in the Cost Sheet is significantly lower than the "Geological Stripping Ratio" (Plan).
    - *Example:* Plan says 1:4 (1 ton ore : 4 tons waste). Accounts show 1:2. Where did the cost of the other 2 tons go? Into the Balance Sheet as an asset.

Fraud 2: The "Cut-Off" Game (Window Dressing)

- The Fraud: Booking April's revenue in March to meet annual targets.
- The Mechanism: Generating Invoices on March 31st (11:59 PM) for material that hasn't left the gate.
- Indicator: A massive spike in "Sales" on the last 3 days of the month, followed by "Sales Returns" or zero dispatch in the first week of April.
- Verification: Match Invoice Date vs. E-Way Bill Date vs. Weighbridge Stamp.

22.8. Management Reporting (MIS) Review

The Finance Controller is the Co-Pilot. The Audit must verify if they are watching the radar.

The Auditor must check if the Monthly MIS Deck highlights:

1. Cost of Production (COP): Trend analysis of Variable vs. Fixed Cost.
2. Breakeven Point: Current sales price vs. COP.
3. Cash Burn Rate: Days of working capital available.
4. Yield Analysis: Reconciliation of Mining Output vs. Processing Input (Losses).

Chapter 23: Treasury & Hedging

This chapter addresses High-Value/High-Risk financial operations where a single error or speculative decision can wipe out the entire operating profit of the mine.

23.1. Process Overview & Objectives

Treasury Management in mining is complex because mining companies are typically Capital Intensive (High Debt) and Net Exporters (High Forex Exposure). The process involves managing liquidity to keep operations running while mitigating the volatility of currency markets and interest rates.

The Value Chain: Cash Flow Forecasting > Fund Arrangement (Debt/PCFC) > Forex Management (Hedging) > Export Realization > Debt Servicing

Process KPIs (Management Dashboard):

- Weighted Average Cost of Capital (WACC): Is the cost of debt optimized?
- Debt Service Coverage Ratio (DSCR): Ability to pay interest + principal.
- Hedge Effectiveness: Did the hedge actually offset the loss?
- Float Time: Average time between export dispatch and cash realization.

23.2. The "Why" Box: Auditor's Mindset

Auditor's Mental Model:

"Operations make the profit; Treasury keeps the company alive.

My biggest fear here is Speculation disguised as Protection.

- The CFO's Bias: CFOs often view the Treasury as a 'Profit Center' rather than a 'Risk Shield.' They might bet on the Dollar falling or Interest rates rising.

- The Risk: If a mine works hard to save 100 per tonne in operations but the Treasury loses 10 Million on a bad currency bet, the company fails.
- The Mandate: Verify that every financial instrument is tied to a real business transaction (Underlying Exposure). If there is a derivative without a cargo, it's gambling."

23.3. Common Industry Jargon

- PCFC (Packing Credit in Foreign Currency): A low-interest loan given to exporters to buy raw materials/fuel before shipment.
- MTM (Mark-to-Market): Valuing a financial asset (like a Forward Contract) at today's market price, not the purchase price.
- Underlying Exposure: The actual business transaction (e.g., an Invoice) that justifies taking a hedge.
- EEFC Account: Exchange Earners' Foreign Currency Account (Allows exporters to hold USD/Foreign currency).
- Covenants: Conditions set by banks (e.g., "Current Ratio must be > 1.33").

23.4. Key Regulatory & Technical References

- Ind AS 109 / IFRS 9: Financial Instruments (Classification, Measurement, and Hedge Accounting).
- Ind AS 21 / IAS 21: The Effects of Changes in Foreign Exchange Rates.
- FEMA 1999: Strict laws against holding foreign currency beyond limits.
- RBI Master Directions on Export of Goods: Guidelines on realizing export proceeds (usually within 9 months).
- Transfer Pricing (Section 92): For inter-company loans and guarantees.

23.5 Detailed Audit Work Program: Treasury & Export Finance

Section 1: Banking & Cash Management

Focus: Prevention of Cash Theft and Liquidity Fraud.

Sub Process	Red Flag / Risk	Audit Procedure (Field Steps)
Old Un-reconciled Entries	Window Dressing: Cheques issued to vendors but kept in the drawer to prevent them from hitting the bank, artificially inflating the Cash Balance on the reporting date.	BRS Scrutiny: Filter the Bank Reconciliation Statement (BRS) for cheques issued > 3 months ago but not cleared. Physically check the "Cheque Outward Register" to see if they were actually handed over to the vendor.
The "Kiting" Scheme	Cash Inflation: Transferring funds between two internal bank accounts (Bank A to Bank B) on March 31st. Bank B records receipt immediately; Bank A records withdrawal 2 days later. Cash is counted twice.	Inter-Bank Trace: Scan all transfers between own accounts on the last 3 days of the month. Match the "Debit Date" in Bank A with the "Credit Date" in Bank B. If they fall in different financial years, reverse the float.
Signatory Matrix Review	Unauthorized Outflow: A single signatory releasing high-value payments in violation of the "Four Eyes" principle (Dual Approval).	Limit Test: Select the top 10 payments of the year. Check the "Payment Advice" sent to the bank. Did two authorized signatories sign it? Does the amount exceed their delegated limit?
Sweep Account Efficiency	Idle Cash Loss: Millions sitting in Current Accounts earning 0%	Interest Audit: If the company uses "Auto-Sweep" (transferring idle cash to Fixed Deposits),

	interest due to treasury laziness or collusion with the bank manager.	verify the interest income. If not, calculate the "Opportunity Loss" of idle funds.
Authorized Signatory Verification	Account Takeover: A fraudster (or rogue employee) submitting a fake Board Resolution to the bank to add themselves or a co-conspirator as an authorized signatory without the company's knowledge.	External Confirmation: Do not rely on internal lists. Request the "Current List of Authorized Signatories" directly from the Bank Relationship Manager. Match this list name-by-name against the approved Board Minutes. Any unknown name is a critical security breach. Also perform above check with Internet user ids
Internet Banking Access Control	Unauthorized Access: Employees who resigned or were transferred to non-finance roles (e.g., Sales) still holding active "Maker/Checker" User IDs for the corporate bank account.	User ID vs. HR Master: Request the "Active User List" from the Bank's Corporate Admin portal. Cross-verify with the HR Employee Master. Deactivate IDs of any user who is no longer in the Treasury/Finance department immediately.
Digital Usage Forensics	Proxy Login / Password Sharing: An authorized "Checker" (CFO) sharing their password with a junior accountant to approve payments. The junior approves their own fraudulent entries.	IP/MAC Address Audit: Request the "Login Activity Log" from the bank portal. Filter for the CFO's User ID. Test: Check the IP Address. If the CFO was on leave/traveling, but the login IP is from the Head Office (or a junior's PC), it proves password sharing or unauthorized access.

Section 2: Export Finance (PCFC & Pre-Shipment)

Focus: Compliance with "End Use" norms and Interest Subvention.

Sub Process	Red Flag / Risk	Audit Procedure (Field Steps)
PCFC End-Use Verification	Fund Diversion: Taking a low-interest PCFC (Packing Credit) loan meant for export inputs (Fuel/Explosives) and diverting it to buy land, pay long-term rupee loans, or invest in the stock market.	Utilization Trace: Select a PCFC disbursement. Trace the money flow. Was it used to pay a valid "Mining Contractor" or "Supplier" bill? Diversion is strictly prohibited by RBI.
PCFC Arbitrage	Regulatory Violation: Taking a PCFC loan at 4% (USD) and investing it in a Rupee Fixed Deposit at 7% to earn risk-free arbitrage. This is "Fund Diversion".	Drawdown vs. Deposit: Compare the dates of PCFC drawdowns with the dates of new Fixed Deposit creation. If they match, flag it as a regulatory breach.
Export Obligation Tracking	Penal Interest Risk: Taking PCFC but failing to export within 360 days. The bank will retroactively charge high penal interest.	Shipping Bill Linkage: Link every specific PCFC loan to a specific "Shipping Bill" (Export proof). Report any loan outstanding > 9 months without shipment.
Interest Subvention	Revenue Leakage: Failing to claim the government subsidy on interest for export credit (if applicable) due to bank error.	Rate Verification: Check the interest charged. Is it the "Card Rate" or the "Subsidized Rate"? If the subsidy is missing, raise a claim.

Section 3: Hedging & Derivatives

Focus: Hedging only against genuine exposure; No Speculation.

Sub Process	Red Flag / Risk	Audit Procedure (Field Steps)
The "Underlying" Test	Speculation (Gambling): The Treasury team betting on currency movements by booking Forward Contracts without any actual Export Order/Invoice to back it up.	Exposure Match: Select 10 open Forward Contracts. Ask: "Show me the Export Order against which this protection was bought." If there is no order, it is speculation.
Mark-to-Market (MTM) Validation	Hidden Losses: The bank reports a small MTM loss, but the reality is a massive loss. Treasury relies solely on the bank's statement.	Independent Valuation: Re-calculate the MTM loss/gain using independent rates (Bloomberg/Reuters) for the closing date. Do not trust the counter-party bank's valuation blindly.
Exotic Derivatives	Unlimited Risk: Entering into "Zero Cost" structures or "Knock-out" options where the company risks infinite loss for a small gain.	Product Suitability: Review the "Deal Ticket." If the product involves "Barriers" or "Leverage," flag it to the Audit Committee. Mining companies should stick to plain vanilla Forwards/Options.

Section 4: Debt Covenants & Compliance

Focus: Preventing Technical Default.

Sub Process	Red Flag / Risk	Audit Procedure (Field Steps)
Covenant Testing	Technical Default: Breaching loan conditions (e.g., Debt/EBITDA ratio, DSCR) which allows the bank to recall the	Ratio Re-Calculation: Calculate the ratios (DSCR, Current Ratio) exactly as defined in the "Sanction Letter." Look for creative accounting used to meet

	entire loan immediately.	these ratios (e.g., treating short-term debt as trade payables).
Security Creation	Legal Breach: Delay in creating a mortgage/charge on mining assets in favor of the lender.	ROC Search: Verify if the charges on assets are registered with the Registrar of Companies (ROC) within the stipulated time (usually 30 days).
Bank Charges Audit	Cost Leakage: Banks overcharging "Processing Fees" or "Forex Conversion Spreads" (e.g., charging 20 paise spread instead of the agreed 5 paise).	Debit Note Check: Pick a sample of Forex remittances. Compare the "Exchange Rate Applied" vs. the "Interbank Rate." Is the spread within the agreed contract limit?.

Section 5: Forensic Checks (Global Fraud Case Studies)

Focus: Detecting "Black Swan" frauds and off-balance-sheet exposures.

Sub Process	Fraud Case Study Reference	Audit Procedure (Field Steps)
Digital Balance Confirmation	The "Wirecard" Check (2020): Wirecard auditors relied on screenshots and documents provided by third-party trustees/management, which were forged. The €1.9 Billion cash did not exist.	The "Blind" Confirmation: 1. Strict Prohibition: Do not accept any PDF statement, email forward, or letter handed over by the Client/Treasury team. 2. Procedure: Use a secure external platform or send a physical letter/encrypted email directly to the Bank's central audit dispatch (not the Relationship Manager). Test: If the bank ignores the auditor but replies instantly to the CFO, suspect collusion.

SWIFT vs. Ledger Reconciliation	The "Nirav Modi / PNB" Check (2018): Bank officials issued "Letters of Undertaking" (Guarantees) via the SWIFT messaging system (MT 760/799) to overseas banks. These were *never entered* in the bank's Core Banking System (CBS) or the company's books, keeping the liability hidden.	The "Message Log" Audit: 1. Source: Request the raw SWIFT / SFMS Outward Message Log directly from the Bank's Trade Finance desk (bypass the company Treasury). 2. Test: Reconcile every "MT 760" (Guarantee) and "MT 700" (LC) in the log against the "Contingent Liability Register" in the Balance Sheet. Red Flag: Any SWIFT message found in the bank log but missing from the company books is a confirmed off-balance-sheet fraud.
Hidden Overdraft Search	The "Patisserie Valerie" Check (2018): The company reported a "Net Cash" position of £28m. In reality, they had secret overdrafts of £10m in subsidiary accounts that were not disclosed to auditors.	Credit Bureau Search: 1. Tool: Pull a Commercial Credit Report (CIBIL / Equifax / Experian) for the company and its subsidiaries. 2. Test: The report lists all active credit lines linked to the company's PAN/Tax ID. Compare this list with the "Bank GL Codes" in the Trial Balance. Red Flag: If the Credit Report shows a loan from "Bank X" but "Bank X" is not in the Trial Balance, management is hiding debt.
Vendor Change Verification	The "Facebook/Google BEC" Check (2013-15): Hackers (or insiders) impersonating a valid vendor (e.g., Quanta Computer) requested a change in bank account details via email. The Treasury wired	The "Call-Back" Protocol: 1. Sample: Filter the "Vendor Master Audit Trail" for changes to Bank Account Numbers for top 10 vendors. 2. Test: Check the supporting evidence. Was there a Voice Confirmation (Call-Back) recorded? 3. Forensic: Verify the IP address of the user who made the change. If the IP maps to a

	$100M+ to the hacker's account.	VPN or non-corporate location, it is likely an Account Takeover.
Investment Valuation	The "IL&FS / Silicon Valley Bank" Check: Holding long-term illiquid assets but valuing them at "Cost" or "Face Value" while the market value has crashed, hiding insolvency.	Distress Testing: 1. Sample: Select Treasury Investments in Bonds/Mutual Funds. 2. Test: Do not rely on the Fund Manager's statement. Check the latest rating of the underlying paper. Red Flag: If a bond is rated "D" (Default) or "C" (Junk) but is still carried at 100% value in the books, insist on an immediate Impairment Provision.

23.7 Management Reporting (MIS) Review

The Auditor must ensure the Finance Controller monitors the following:

1. Liquidity Runway: (Cash + Undrawn Lines) / Monthly Burn. ("How many months can we survive without sales?")
2. Open FX Position: Net Unhedged Exposure (USD Millions).
3. Cost of Debt: Weighted Average Interest Rate trend.
4. Forex Realization Aging: Are any export proceeds pending > 9 months? (Regulatory Risk).

23.8. Common Audit Observations

1. Lazy Treasury: Keeping excess funds in the EEFC (Dollar) account when the Rupee is appreciating, resulting in FX losses on translation.
2. Covenant Blindness: The Finance team doesn't track Bank Covenants monthly. They only realize they are in breach at year-end when the auditor points it out.

3. Bank Charges Leakage: Banks often overcharge "Processing Fees" or "Forex Conversion Charges" (e.g., charging 20 paise spread instead of the agreed 5 paise). Treasury rarely checks the debit notes.

23.9. Key Frauds & Indicators

Fraud 1: The "Kiting" Scheme (Window Dressing)

- The Concept: Inflating the Cash Balance on the Balance Sheet date.
- The Mechanism:
    - Company has two accounts: Bank A and Bank B.
    - On March 31st, a cheque is drawn on Bank A and deposited in Bank B.
    - Bank B records the receipt immediately (Cash increases).
    - Bank A records the withdrawal 2 days later (Float).
    - Result: The cash is counted twice on March 31st.
- Indicator: High value transfers between own accounts on the last day of the month.

Fraud 2: The "Tex-Mex" Derivative Disaster

- The Concept: Entering into complex, high-risk derivatives to show "Treasury Income."
- The Mechanism: The CFO buys a "Call Spread" or "Barrier Option." If the currency stays stable, they earn a premium (Profit). If the currency crashes, they owe the bank millions.
- Indicator: The Treasury P&L shows steady small profits every month, disconnected from actual export volume. This is the "calm before the storm."

Mine Development: This is the heavyweight chapter of the Balance Sheet. In mining, the distinction between "Operating Cost" (moving dirt) and "Capital Asset" (moving dirt to access *future* ore) is often a multimillion-dollar accounting judgment.

24.1. Process Overview & Objectives

Fixed Asset management in mining goes beyond buying machinery. It involves the capitalization of "Pre-stripping" (Earthworks), "Mine Development" (Infrastructure), and huge yellow metal fleets. The objective is to accurately reflect the Earning Capacity of the mine while matching the cost of the asset against the revenue it generates over its life (Depreciation).

The Value Chain: Capex Budgeting > Procurement (Chapter 14) > Capitalization (CWIP to Gross Block) > Depreciation/Amortization > Impairment > Disposal/Scrap

Process KPIs (Management Dashboard):

- Capex Budget Variance: Actual spend vs. Budget (Cost Overrun).
- Asset Turnover Ratio: Revenue / Total Assets.
- ROCE (Return on Capital Employed): EBIT / Capital Employed.
- Stripping Ratio Variance: Actual SR vs. Life-of-Mine SR (Impacts Capitalization).

24.2. The "Why" Box: Auditor's Mindset

Auditor's Mental Model:

"Assets are the graveyard of buried expenses.

My primary suspicion is Aggressive Capitalization.

- The Manager's Motivation: If I repair a gearbox for 500000 and expense it, my monthly profit drops. If I 'Capitalize' it, the cost is spread over 5 years. My bonus is safe today.

- The Risk: The Balance Sheet fills up with 'Assets' that are actually just repairs or waste removal expenses.
- The Counter-Risk: Ghost Assets. We are paying insurance and depreciation on machines that were stolen or scrapped years ago."

24.3. Common Industry Jargon

- CWIP (Capital Work in Progress): Assets under construction/installation.
- Gross Block: Historical cost of assets. Net Block: Cost minus accumulated depreciation.
- Stripping Activity Asset: The capitalized value of waste removal (Overburden) during the production phase (per IFRIC 20).
- Componentization: Breaking a large asset (e.g., Excavator) into smaller parts (Bucket, Engine, Chassis) for separate depreciation.
- Impairment: A permanent reduction in the value of an asset (e.g., when a mine floods).

24.4. Key Regulatory & Technical References

- Ind AS 16 / IAS 16: Property, Plant & Equipment (The Rulebook).
- IFRIC 20: Stripping Costs in the Production Phase of a Surface Mine (The Mining Bible).
- Ind AS 36: Impairment of Assets.
- Ind AS 23: Borrowing Costs (Capitalizing interest during construction).
- CARO 2020 (Clause i): Auditor *must* report on Physical Verification discrepancies and Title Deeds.
- GST Act (Sec 17(5)): Input Tax Credit not allowed for civil structures (buildings), but allowed for Plant & Machinery.

24.5 Detailed Audit Work Program: Fixed Assets & Capitalization

Section 1: Capitalization & CWIP Management

Focus: Ensuring only valid assets are capitalized and projects are closed timely.

Sub Process	Red Flag / Risk	Audit Procedure (Field Steps)
Initial Spares (Bundled Spares)	Classification Error: High-value "Insurance Spares" supplied with the main machine are dumped into the "Consumables Store" (Inventory) instead of being capitalized, or vice versa (capitalizing filters/gaskets).	PO vs. Booking Check: Review the Purchase Order of major machinery (e.g., Crusher, HEMM). Test: Identify line items for "Commissioning Spares." 1. Consumables (Oils/Filters): Must be expensed/inventoried. 2. Critical Spares (Rotors/Motors): Must be capitalized as part of the asset or as a separate "Capital Spare" if life > 1 year.
Refurbished Assets (Derecognition)	Double Counting: Capitalizing the cost of a "Refurbishment" (e.g., New Engine for Dumper) but failing to write off the Book Value of the *old* replaced engine. The asset value becomes inflated (Old Part + New Part).	The "Substitution" Test: Select 5 major refurbishment capitalizations. Test: Check the Fixed Asset Register (FAR). Was the "Gross Block" of the *old* replaced component removed/adjusted? If not, the asset is overstated, and depreciation is incorrect.

Repairs vs. Capitalization	Profit Inflation: Booking routine repairs (e.g., "Annual Maintenance") as "Capital Additions" to reduce current Opex.	Voucher Scrutiny: Scan "Plant Additions" for keywords like "Service," "Overhaul," or "AMC." Capitalize *only* if the expense increases capacity or extends useful life beyond original specs.
Interest Capitalization	Cost Inflation: Capitalizing interest cost *after* the asset is ready for use.	"Ready for Use" Test: Check the "Commissioning Certificate" date. Ensure interest capitalization stopped exactly on that date.

Section 2: Asset Master Data & Maintenance Linkage (New)

Focus: Bridging the gap between Finance (FAR) and Engineering (PM).

Sub Process	Red Flag / Risk	Audit Procedure (Field Steps)
FAR vs. Equipment Master (The "Orphan" Asset)	Maintenance Gap: An asset exists in the Finance Books (FAR) but is missing from the Plant Maintenance (PM) Master. Consequence: No maintenance schedules (Lubrication/Service) are generated for it, leading to premature breakdown.	Two-Way Reconciliation: 1. FAR to PM: Pick 20 assets from the Finance Register. Search for them in the SAP PM / other PM system. 2. PM to FAR: Pick 20 active machines from the Maintenance Schedule. Check if they exist in the Finance Books. Result: Mismatches indicate "Ghost Assets" or "Unmaintained Assets."
Component Tagging	Tracking Failure: The Finance Asset ID is stuck on the "Main Chassis," but the	Rotable Asset Audit: Check the "Component Tracking Log" in the Workshop.

	Engine and Transmission (rotable items) are swapped between trucks. Finance continues depreciating the "Old Engine" on Truck A, while it is physically running on Truck B.	Test: Pick an engine Serial Number physically installed on Dumper #101. Does the Finance Register reflect this engine as part of Dumper #101? If not, the component-level depreciation data is junk.

Section 3: Mine Development (Stripping Costs- IFRIC 20)

Focus: Correct treatment of Overburden (Waste) Removal costs.

Sub Process	Red Flag / Risk	Audit Procedure (Field Steps)
Stripping Ratio (SR) Test	Aggressive Capitalization: Capitalizing routine waste removal cost when the stripping ratio is normal.	Ratio Comparison: Compare Actual Stripping Ratio vs. Standard Stripping Ratio. Capitalization is allowed *only* if Actual SR > Standard SR (Accessing future ore).
Identifiable Component	Vague Capitalization: Capitalizing "General Waste Removal" without linking it to a specific future ore body.	Geologist Interview: Ask: "Which specific ore body section is this stripping accessing?" If they cannot identify a distinct component, capitalization is invalid.

Section 4: Physical Verification & Disposal

Focus: Asset existence and safeguarding.

Sub Process	Red Flag / Risk	Audit Procedure (Field Steps)
Floor-to-Book Check	Unrecorded Assets: Machines working in	Site Tag Check: Select 20 machines physically working in

	the pit but not found in the books.	the mine. Trace their Tag/Serial Numbers back to the FAR.
Book-to-Floor Check	Ghost Assets: Machines sold/scrapped years ago still appearing in the FAR.	Existence Check: Select 20 high-value assets from FAR. Physically find them. If missing, investigate for theft.
Refurbishment WIP Control	Theft of Old Parts: When an asset is sent outside for refurbishment, the old valuable parts (Copper coils, heavy steel) are stolen by the vendor instead of being returned/credited.	Gate Pass Return Audit: Check the "Returnable Gate Pass" for assets sent for repair. Test: When the refurbished asset returned, did the vendor return the *scrap/old parts*? Check the "Scrap Inward Register." If not, the scrap value was leaked.

24.6. Common Audit Observations

1. The "Forever" CWIP: A conveyor belt project was abandoned 4 years ago due to land disputes. The cost is still sitting in CWIP. Management refuses to write it off to avoid hitting the P&L.
2. Tyre Capitalization: Buying a set of OTR Tyres and capitalizing them as "Plant & Machinery" with a 15-year life. Tyres are consumables (Opex) or 1-year assets.
3. GST Input Credit Risk: Claiming GST credit on the construction of the "Admin Building" or "Staff Colony." This is blocked under Sec 17(5) and attracts 18% interest + penalty.

24.7. Key Frauds & Indicators

Fraud 1: The "Gold Plated" Asset (Procurement Fraud)

- The Concept: Buying a premium machine on paper but receiving a standard machine physically.
- The Mechanism:
  - PO Specs: "Excavator with GPS, Auto-Lube, and AC Cabin." Price: 3000k.
  - Physical Asset: "Standard Excavator." Price: 2500k.
  - The 500k difference is split between Vendor and Purchase Manager.
- Indicator: Physical Verification reveals discrepancies between "Nameplate Specs" and "PO Specs."

Fraud 2: The "Ghost" Fleet (Insurance Fraud)

- The Mechanism: A Dumper falls into the pit and is a total loss. Insurance pays the claim.
- The Fraud: The company keeps the wrecked dumper on the Asset Register. They continue to claim depreciation on it. Later, they might even "cannibalize" parts from it while it's still listed as an active asset.
- Indicator: Assets with "Active" status in FAR but "Zero Utilization" in the Fleet Management System (FMS) for > 6 months.

24.8. Management Reporting (MIS) Review

The Finance Controller must monitor:

1. Capex Variance: Budget vs. Actual (Are we overspending?).
2. Asset Utilization %: Running Hours/Available Hours. (Why buy new machines if current ones are idle?).
3. CWIP Ageing: List of projects > 1 year old.
4. Stripping Ratio Trend: Is the capitalization of waste removal sustainable, or are we just pushing costs to the future?

Part VIII: Sustainability, Governance, Risk & Compliance

Chapter 25: ESG Assurance & BRSR Reporting

This chapter marks the shift from "Voluntary Marketing" to "Regulatory Assurance." The auditor's role is to treat Carbon and Safety data with the same rigor as Cash and Revenue.

25.1. Process Overview & Objectives

ESG (Environmental, Social, and Governance) reporting has moved from voluntary "Marketing Brochures" to mandatory Regulatory Filings (like BRSR in India). The objective is to verify the Non-Financial Data. Just as we audit the Balance Sheet for financial truth, we audit the ESG Report for "Green Truth."

The Value Chain: Data Capture (Energy/Water/Safety) > Conversion Factors (Emission Co-efficient) > Consolidation > BRSR/GRI Reporting > External Assurance

Process KPIs (The Sustainability Dashboard):

- GHG Intensity: Co2 per Tonne of Ore produced.
- Water Positivity Ratio: Water Returned vs. Water Consumed.
- LTIFR: Lost Time Injury Frequency Rate (Safety Metric).
- CSR Spend Utilization: Actual Spend / Statutory 2% Requirement.

25.2. The "Why" Box: Auditor's Mindset

Auditor's Mental Model: "My job is to detect Greenwashing".

Companies love to claim 'Net Zero' and 'Water Positive'.

- The Risk: If a mine claims to be 'Water Neutral' but local villagers are protesting about dry wells, the report is a lie.
- The Consequence: In the modern era, a fake ESG report leads to Divestment by global funds (BlackRock/Vanguard) and regulatory penalties (SEBI).
- The Approach: Treat Carbon Data exactly like Cash Data. Trace it to the source (Meter/Bill)."

ESG Data Architecture & Assurance Map: ESG reporting is only as strong as the weakest spreadsheet feeding it. Climate is not just an ESG narrative; it affects mine economics and reserve life.

- Source-to-Disclosure Mapping
    - Audit Check: Map each key ESG metric (GHG Scope 1 & 2, water withdrawal, waste, injuries, diversity, community spend) to its source systems: FMS, weighbridge, CPP DCS, HRMS, EHS software or manual logs.
    - Risk: KPIs "manufactured" in Excel files with no version control or audit trail.
- Assurance Layering
    - Audit Check: Prepare an assurance matrix showing which metrics are covered by internal audit, statutory audit, external ESG assurance or remain unaudited.
    - Red Flag: High-visibility metrics (like Scope 1 emissions) with no independent assurance despite being used in targets/bonuses.
- Physical & Transition Risk Assessment
    - Audit Check: Review any climate scenario/physical risk studies (heat, flooding, water scarcity) and confirm if outcomes are reflected in LoM planning, capex (e.g., flood defences) and insurance.
    - Risk: Mines in flood-prone areas without any capex for drainage despite repeated events.
- Decarbonisation Roadmap & Tracking
    - Audit Check: Assess targets for diesel, grid power and explosives emissions; verify project lists (solar/wind, trolley-assist, IPCC, electrified HEMM) and status tracking.
    - Red Flag: Public "Net-Zero" declarations with no internal project pipeline or budgets.

25.3. Common Industry Jargon

- BRSR: Business Responsibility and Sustainability Report (Mandatory for Top 1000 listed co's in India).
- Scope 1 Emissions: Direct emissions from sources owned (e.g., Diesel burnt in Dumpers).
- Scope 2 Emissions: Indirect emissions from purchased electricity (e.g., Grid Power).
- Scope 3 Emissions: All other emissions (e.g., Transport by contractors, Employee commute, Use of sold coal).
- Materiality Assessment: The process of identifying which ESG issues matter most to *this* specific company.
- Greenwashing: Exaggerating green credentials.

25.4. Understanding the Emission Sources (The Carbon Audit)

- Scope 1 (Direct Emissions): Emissions from sources that are owned or controlled by the mining company.
    - *Mining Examples:* Diesel burned in Dumpers/Excavators, Coal burned in the Captive Power Plant (if owned), Blasting fumes (Explosives).
- Scope 2 (Indirect Energy Emissions): Emissions from the generation of purchased electricity consumed by the company.
    - *Mining Examples:* Electricity bought from the State Grid to run the Crusher, Mill, or Office lighting.
- Scope 3 (Other Indirect Emissions): All other emissions that occur in the value chain (upstream and downstream).
    - *Mining Examples:* Emissions from ships/trains transporting ore to customers (Downstream), Employee commuting in buses, Cement production for mine construction (Upstream).

25.5. Key Regulatory & Technical References

- Regulations (Indian/Global):
  - SEBI LODR (Reg 34): Mandates BRSR filing.
  - Companies Act (Sec 135): CSR Spending (2% of Avg Net Profit).
- Standards:
  - GRI Standards: Global Reporting Initiative (The Accounting Standard for ESG).
  - GHG Protocol: How to calculate Carbon Footprint.
  - SASB: Industry-specific standards (Metals & Mining).

25.6 Detailed Audit Work Program: ESG Assurance & BRSR

Section 1: Environmental Assurance (The "Green Truth")

Focus: Verifying Carbon, Energy, and Water claims.

Sub Process	Red Flag / Risk	Audit Procedure (Field Steps)
Scope 1 (Direct Emissions)	The "Contractor" Exclusion: Reporting lower emissions by excluding diesel consumed by contractor's dumpers working inside the mine. Under GHG protocol, if you control the operation, it is Scope 1.	Fuel Reconciliation: Calculate Total Diesel Consumption (Company + Contractor). Multiply by the Emission Factor. Compare this with the "Scope 1" figure in the BRSR report.

Scope 2 (Indirect Energy)	Double Counting: Selling the "Green Attribute" (Carbon Credit/REC) to a third party but still claiming "Zero Carbon" or "Renewable Usage" in the Sustainability Report.	REC Inventory Check: If the company claims "100% Renewable Power," check the Renewable Energy Certificates (REC) registry. Were these certificates *retired* in the company's name or *sold* on the exchange? You cannot sell the credit and claim the green benefit simultaneously.
Water Positivity (ZLD)	The "Phantom" Recycle: Claiming "Zero Liquid Discharge" (ZLD) status while flow meters show zero recycling water returning to the process.	Water Balance Audit: Compare "Water Withdrawal" (Input Flow Meter) vs. "Water Discharge" (Output). If Withdrawal is high but Recycling is near zero, the water is being dumped illegally.
Decarbonization Roadmap	Greenwashing: Public "Net Zero" declarations with no internal project pipeline or budget allocated to achieve them.	Capex Linkage: Review the approved Capex Budget. Are there specific line items for "Solar Plants," "Electric Fleet," or "Trolley Assist"? If not, the Net Zero pledge is a marketing lie.

Section 2: Social Assurance (Safety & Human Rights)

Focus: Honest reporting of workplace reality.

Sub Process	Red Flag / Risk	Audit Procedure (Field Steps)
LTIFR Integrity (Safety)	Suppression of Accidents: A worker	Clinic vs. HR Register: Compare the "Medical Clinic Register"

	breaks a finger (Lost Time Injury), but is given "Light Duty" in the office so the mine can report "Zero LTI".	(First Aid Log) with the "Official Accident Report" sent to DGMS. Look for "Sick Leave" applications on dates matching Clinic entries for injuries.
R&R (Resettlement)	False Rehabilitation: Reporting "100% Rehabilitation" of displaced families when they are still living in temporary shelters 5 years later.	Physical Census: Select 10 Project Affected Families (PAFs) from the "Settled" list. Visit their relocation site. Do they have the house/job promised in the R&R Policy?.
Diversity Reporting	Headcount Inflation: Counting contract labor women (sweepers/gardeners) to inflate the "Women in Workforce %" for the Annual Report.	Payroll Segregation: Verify the count of female employees in the Permanent Payroll. Ensure the definition of "Workforce" in the report clearly distinguishes between Permanent and Contract staff.

Section 3: Data Architecture & Governance

Focus: Is the data reliable or "manufactured" in Excel?

Sub Process	Red Flag / Risk	Audit Procedure (Field Steps)
Source-to-Disclosure	Excel Fatigue: KPIs "manufactured" in manual Excel files with no version control or audit trail, leading to data manipulation.	Trace to Source: Pick one metric (e.g., Energy Intensity). Trace it back from the Annual Report > ESG Dashboard > Excel Sheet > Electricity Bill/Meter. Map the flow.
Whistleblower Mechanism	Conflict of Interest: The Whistleblower	"Mystery Email" Test: Send a dummy complaint to the

	email goes directly to the HR Head or Legal Head (who might be the perpetrator) instead of the Audit Committee Chair.	whistleblower hotline/email. Track who receives it and the response time. Ensure it bypasses executive management.
Supplier Code of Conduct	Scope 3 Risk: Suppliers using Child Labor or violating environmental laws, creating reputational risk for the mine.	Vendor Audit: Check the top 10 suppliers (Explosives/Transport). Did they sign the ESG Code of Conduct? Has the company conducted a "Second Party Audit" on them?.

25.7. Common Audit Observations

1. The "Estimates" Game: Water consumption reported based on "Pump Capacity x Hours" instead of Flow Meters. This is usually 30% inaccurate.
2. Scope 3 Blindness: The mine reports low emissions but ignores the massive emissions from transporting the ore (Rail/Ship) to the customer.
3. CSR "Donation" vs. "Impact": Spending CSR funds on "General Donations" to local political festivals instead of sustainable projects (Schools/Hospitals).

25.8. Key Frauds & Indicators

Fraud 1: The "Greenwashing" of Scope

- The Concept: Companies choose boundaries that make them look good.

- The Fraud: A mining company spins off its "Captive Power Plant" (Thermal) into a separate subsidiary. It then claims the Mine has reduced its emissions by 80%.
- Indicator: A massive drop in Scope 1 emissions accompanied by a massive rise in "Purchased Power" or Related Party Transactions.

Fraud 2: The "Safety Triangle" Manipulation

- The Concept: Heinrich's Triangle says for every 1 Major Injury, there are 29 Minor Injuries and 300 Near Misses.
- The Fraud: The report shows Zero Near Misses and Zero Minor Injuries, but 1 Fatality.
- Indicator: This statistical impossibility proves that the safety reporting culture is suppressed/fear-based.

25.9. Management Reporting (MIS) Review

The Sustainability Officer must present to the Audit Committee:

1. Energy Intensity Trend: kWh/Tonne. (Is efficiency improving?)
2. Non-Compliance Tally: Number of "Show Cause Notices" from Pollution Control Board.
3. Grievance Closure Rate: How many community complaints (dust/noise) are open > 30 days?
4. CSR Impact Assessment: Third-party verification of where the money went.

This area is critical because a single violation of Environmental or Safety laws can lead to immediate mine closure or criminal prosecution of the Occupier/Manager.

26.1. Process Overview & Objectives

Mining is one of the most heavily regulated industries in the world. The objective of this audit is to ensure the mine operates within the legal boundaries set by the Directorate General of Mines Safety (DGMS) and the Pollution Control Board (PCB).

The Compliance Value Chain:

Regulatory Application (EC/CTO) > Operational Implementation > Monitoring (Air/Water/Safety) > Statutory Return Filing > External Inspection

Process KPIs (Management Dashboard):

- Compliance Score: % of statutory conditions met.
- Show Cause Notices (SCN): Number of notices received from regulators.
- LTIFR: Lost Time Injury Frequency Rate.
- Discharge Parameters: Variance between Actual vs. Permissible limits (SPM, pH, COD).

26.2. The "Why" Box: Auditor's Mindset

Auditor's Mental Model:

"I am not an Environmental Engineer, nor a Safety Officer. But I am the Risk Assessor.

- The Risk: If the 'Consent to Operate' (CTO) expires and we keep mining, the revenue generated is Illegal Proceeds. The government can confiscate 100% of it (Odisha Mining Case precedent).
- The Blind Spot: Operations teams often hide 'Near Misses' and 'Pollution Spikes' to protect their bonuses.

- The Goal: Confirm that the 'Paper Compliance' (Registers) matches the 'Ground Reality' (Physical Site)."

26.3. Common Industry Jargon

- EC (Environmental Clearance): The master permission from the Ministry (MoEFCC) with specific conditions (e.g., "Plant 10,000 trees").
- CTO (Consent to Operate): Periodic license from the State Pollution Control Board (Water & Air Acts).
- DGMS: Directorate General of Mines Safety (The Police of Mining Safety).
- PESO: Petroleum and Explosives Safety Organization (Regulates the Magazine).
- Form-A / Form-B: Employee registers required under Mines Rules.
- VT (Vocational Training): Mandatory training for every worker before entering the mine.

26.4. Key Regulatory & Technical References

- Safety Statutes (Indian Context):
    - The Mines Act, 1952: The Constitution of mining.
    - Mines Rules, 1955: Rules for welfare (Water, Rest Shelter, Medical).
    - CMR 2017 (Coal Mines Regs) / MMR 1961 (Metalliferous Mines Regs): Technical safety rules.
- Environment Statutes:
    - Environment (Protection) Act, 1986.
    - Air (Prevention and Control of Pollution) Act, 1981.
    - Hazardous Waste Management Rules, 2016.

26.5 Detailed Audit Work Program: Statutory & Industrial Compliance

Section 1: Environmental Compliance (EC & CTO)

Focus: Adherence to Environmental Clearance (EC) and Consent to Operate (CTO) conditions.

Sub Process	Red Flag / Risk	Audit Procedure (Field Steps)
The "Conditions" Audit	Paper Compliance: The file says "Complied," but the physical infrastructure (e.g., Garland Drain) is broken, silted up, or non-existent.	Physical Verification: Take the Environmental Clearance (EC) document which lists specific conditions (e.g., "Construct a Garland Drain around the dump"). Go to the site. Does the drain exist and is it functional?
Stack/Ambient Monitoring	The "Sunday Sample": Pollution samples consistently taken on days when the crusher was shut down or operating at 50% capacity to show low pollution results.	Lab Report Review: Review the Third-Party Lab reports for Air Quality (PM 10, PM 2.5). Correlate the sample dates with the Production Logbook. If samples coincide with low production days, the data is manipulated.
Zero Liquid Discharge (ZLD)	Illegal Discharge: Hidden bypass pipes leading to nearby rivers/drains to dump untreated water and save treatment costs.	ETP Audit: Verify Effluent Treatment Plant (ETP) logbooks. Check if the "Outlet Valve" is physically sealed. Inspect the perimeter for unauthorized pipes exiting the boundary wall.
Expansion Audit	Illegal Mining: Constructing new infrastructure (e.g., a New Conveyor Line) without obtaining	Capex Reconciliation: Compare the "Fixed Asset Register" additions with the CTE files. If a new facility exists physically but has no CTE, it is an illegal asset.

	"Consent to Establish" (CTE).	

Section 2: Mines Safety (DGMS Compliance)

Focus: Prevention of accidents and legal immunity for management.

Sub Process	Red Flag / Risk	Audit Procedure (Field Steps)
Statutory Manpower Count	Shortage of Overmen: Running 3 shifts with only enough statutory supervisors for 1 shift. This is a direct violation of MMR/CMR and risks safety.	Roster Verification: Compare the "Required Statutory Staff" (based on Output/Manpower as per Act) vs. "Actual Appointed" staff on the shift roster.
Vocational Training (VT)	Untrained Workers: Workers deployed without mandatory safety training. The VT Centre often issues certificates without conducting actual classes.	Spot Check: Randomly pick 5 contract workers in the pit. Ask for their "VT Card." verify the dates and quiz them on basic safety signs.
PME (Medical Exam)	Unfit Workers: Expired medical certificates leading to unfit workers operating heavy machinery.	Form O Check: Check Form O. Has every worker >45 years undergone a medical checkup every 3 years? Verify the dates against the current roster.
Safety Management System (SMS)	Paper Safety: Risk assessments (HIRA) and Standard Operating Procedures (SOPs) existing only on paper with identical	Toolbox Talk Audit: Attend a morning "Toolbox Talk." Is it relevant to the day's specific risks (e.g., rain, blasting), or just a generic speech? Check if the Risk

	text copied across months.	Register is updated for new hazards.
Near Miss Reporting	Suppression: The safety report shows "Zero Near Misses," which is statistically impossible and indicates a fear culture.	Leading Indicators: Review the "Near Miss" or "Safety Observation" cards. If there are none, or if contractors are excluded from reporting, the safety culture is broken.

Section 3: Hazardous Waste & Explosives

Focus: Safe handling of dangerous materials.

Sub Process	Red Flag / Risk	Audit Procedure (Field Steps)
Used Oil Disposal	Illegal Selling: Selling used oil to local brick kilns (illegal) instead of authorized recyclers.	Manifest Audit: Track the "Used Oil" generated from the workshop. Where did it go? Check the "Manifest Form 10" to ensure it was sent to a registered recycler.
Magazine Audit	Explosion Risk: Poor maintenance of the explosive magazine, such as high temperatures or faulty lightning protection.	Logbook Check: Check the "Lightning Conductor" testing record and the "Temperature Log" of the explosive magazine. Ensure temperature is maintained < 30°C.

26.6. Common Audit Observations

1. The "Ghost" Safety Committee: The Mines Rules require a Safety Committee to meet monthly. The minutes are often copy-pasted from the previous month, with the same attendees (some of whom were on leave).

2. Garland Drain Neglect: The EC requires concrete drains to prevent runoff. Auditors often find these drains choked with mud, rendering them useless during monsoons.
3. Dust Suppression Failure: The logbook shows "Water Sprinkler: Active 8 Hours." The GPS shows the water truck was parked at the canteen for 4 hours.

26.7. Key Frauds & Indicators

Fraud 1: The "Certificate Renting" Scam (Statutory Manpower)

- The Concept: Mines need a "First Class Mines Manager" by law. They are expensive and scarce.
- The Fraud: The mine hires a Manager *on paper*. He submits his certificate but lives in another city. A junior, unqualified person runs the mine using the Manager's signature stamp.
- Indicator: Check the Biometric/Gate attendance of the Mines Manager. Does he physically enter the mine daily? Compare his signature on the "Shift Report" with his signature on the "Leave Application."

Fraud 2: The "Sample Management" (Pollution Control)

- The Concept: Pollution levels must be within limits.
- The Fraud: The Environment Officer instructs the lab to take air samples *only* after heavy rain (when dust is settled) or positions the equipment *upwind* of the crusher.
- Indicator: Air Quality results that are consistently 50% lower than the industry average or show zero variance day-to-day.

26.8. Management Reporting (MIS) Review

The Auditor must ensure the Board is seeing the "Red" flags, not just Green.

The Compliance Dashboard must track:

1. License Expiry Radar: List of licenses expiring in the next 90 days (EC, CTO, Factory License).
2. Statutory Vacancy: % of mandatory positions (Manager/Foreman) vacant.
3. Non-Compliance Aging: SCNs (Show Cause Notices) open for > 6 months.
4. Safety Budget Utilization: Is the budget being spent on PPE/Training, or is it lapsing?

SME Note: The "Deemed Owner" Risk

Under the Mines Act, if the Owner (Board of Directors) does not provide financial resources for safety, they are criminally liable for accidents. The Auditor must verify that "Safety Budget Requests" from the Mine Manager were never rejected by the Finance Department.

Chapter 27: Secretarial, CSR & Legal

This chapter covers the "Social License to Operate." Mining requires massive land acquisition and affects local communities. This creates complex legal obligations (R&R) and social obligations (CSR). If these are managed corruptly, the mine faces protests, blockades, and legal shutdowns.

27.1. Process Overview & Objectives

This chapter audits the Governance (Secretarial), the Social Spend (CSR), and the Land Acquisition process.

- Land: Acquiring land in India is a legal minefield involving "Right of Way," "Forest Rights Act," and "Rehabilitation & Resettlement (R&R)."
- CSR: It is no longer charity; it is a statutory obligation (Companies Act, 2013).
- Secretarial: Ensuring the Board of Directors is protected from "Deemed Owner" liability.

The Value Chain: Land Identification > Social Impact Assessment (SIA) > Land Acquisition/Lease > R&R Payouts > CSR Project Execution > Secretarial Compliance

Process KPIs (Management Dashboard):

- Land Possession: Physical possession vs. Paper possession.
- R&R Settlement Ratio: Families settled vs. Total Project Affected Families (PAFs).
- CSR Spend Utilization: Actual Spend vs. Statutory Target (2%).
- Litigation Contingency: Value of lawsuits against the company.

27.2. The "Why" Box: Auditor's Mindset

Auditor’s Mental Model:

"This is where the 'Soft Cash' flows.

- The Risk: Buying land involves dealing with local aggregators, farmers, and politicians. It is the easiest place to hide bribes (inflated land rates).
- The CSR Trap: CSR funds are often diverted to 'ghost NGOs' or political favorites to buy peace.
- The Goal: Ensure that every Rupee spent on Land/CSR has a visible, legal trail and actually reaches the intended beneficiary, not a middleman."

27.3. Common Industry Jargon

- R&R (Rehabilitation & Resettlement): The compensation package (House + Job + Cash) given to displaced families.
- PAF (Project Affected Family): A family losing land/livelihood.
- SIA (Social Impact Assessment): A mandatory study to determine the social cost of the mine.
- CSR (Corporate Social Responsibility): Mandatory social spending (Sec 135).
- Encumbrance: A legal claim (mortgage/lien) on land.
- Deemed Owner: The Director/Agent nominated under the Mines Act to take legal responsibility.

27.4. Key Regulatory & Technical References

- LARR Act 2013: The Right to Fair Compensation and Transparency in Land Acquisition, Rehabilitation and Resettlement Act.
- Companies Act, 2013 (Sec 135): CSR Rules.
- Forest Rights Act (FRA), 2006: Rights of tribal communities.
- PMLA (Prevention of Money Laundering Act): Land transactions are high-risk for KYC.

27.5 Detailed Audit Work Program: Secretarial, CSR & Legal

Section 1: Land Acquisition & R&R (The "Soft Cash" Risk)

Focus: Ensuring genuine payments to genuine landowners.

Sub Process	Red Flag / Risk	Audit Procedure (Field Steps)
Land Aggregator Payments	The "Mark-up" Scam: The Aggregator buys land from farmers at ₹1 Lakh/acre and sells it to the company at ₹5 Lakhs/acre the next day. The margin is kicked back to officials.	Deed Value Comparison: Review payments made to Land Aggregators. Compare the Registered Sale Deed Value (amount paid to the farmer) vs. the Company Payout (amount paid to the aggregator). Large variances without value-add justification are red flags.
R&R Beneficiary Verification	Ghost PAFs: Monthly "Subsistence Allowance" or compensation flowing to bank accounts of people who never lived in the village (fake IDs created by local strongmen).	Physical Census: Select 10 Project Affected Families (PAFs) receiving monthly cash benefits. Physically verify their existence and residence history in the village. Match their ID proofs with the original SIA (Social Impact Assessment) survey list.
The "Structure" Fraud	Compensation Inflation: Locals building temporary mud huts or digging fake wells *overnight* after the acquisition notification to claim "Structure	Satellite Forensics: Compare current claim photos with Google Earth/Satellite Imagery from 6 months prior. If the land was empty in Jan and had 50 houses in Feb (just before the survey), the structures are fake.

	Compensation" in addition to land value.	
Job for Land Scheme	Nepotism: Providing permanent jobs to individuals who have no blood relation to the displaced land-owner, often based on union recommendations.	Genealogy Check: Verify the "Family Tree" certificate issued by the local administration. Does the job recipient actually belong to the PAF's immediate family?.
GPS Boundary Audit	Encroachment Risk: Mining outside the acquired land area, leading to "Illegal Mining" penalties (Sec 21(5) MMDR Act).	Overlay Test: Overlay the "Monthly Survey File" (Shapefile) on the "Acquired Land Map" in GIS software. If excavation polygons cross the boundary line, immediate "Stop Work" is required.

Section 2: CSR Management (Statutory Compliance)

Focus: Impactful spending vs. Fund diversion.

Sub Process	Red Flag / Risk	Audit Procedure (Field Steps)
NGO Due Diligence	The "Briefcase NGO": CSR funds diverted to an NGO formed just 1 month before the contract, operating from a residential address, often owned by a politically exposed person (PEP).	Form CSR-1 Check: Verify the "Form CSR-1" registration and the track record of the implementation agency. Visit their registered office to ensure it is a functioning entity, not a shell.
Asset Verification	Multiple Invoicing: The same school building or solar light project	Physical Tagging: Visit the CSR project site (e.g., School/Community Center). Is

	shown to three different companies (e.g., Tata, Adani, Vedanta) as *their* CSR project.	there a permanent plaque or painting indicating *your* company's sponsorship? Check GPS geotags of the construction photos.
"General Donation"	Ineligible Spend: Booking large sums as CSR under "Promotion of Culture," which are actually donations to local political festivals (Dussehra/Diwali) to appease local strongmen.	Expense Scrutiny: Review the ledger for "Donations" or "Sponsorships." If the money didn't create a long-term asset or program, it may be disallowed under the Companies Act (Sec 135) as it is not a scheduled CSR activity.
Impact Assessment	Boilerplate Reporting: Relying on generic "Impact Reports" that say "Everything is excellent" without data support.	Third-Party Review: Read the Social Impact Assessment Report (mandatory for projects > ₹1 Cr). Does it quantify the outcome (e.g., "Literacy increased by 5%") or just list activities?.

Section 3: Secretarial & Legal (Companies Act Checks)

Focus: Governance and Director Liability.

Sub Process	Red Flag / Risk	Audit Procedure (Field Steps)
Related Party Transactions (RPT)	Arm's Length Violation: Contracts awarded to Directors' relatives or sister concerns without Audit Committee approval.	Register Check: Review the "Register of Contracts" (MBP-4). Cross-verify with the Vendor Master. Are there vendors with addresses matching Directors' addresses? Ensure all RPTs are at "Arm's Length Pricing".

Litigation Provisioning	Liability Suppression: A lawyer advises "10% chance of winning," but the company discloses it as "Contingent Liability" (Footnote) instead of creating a "Provision" (Expense) in the P&L.	Legal Opinion Review: Review the "Legal Opinions" file for pending cases. Compare the lawyer's probability assessment with the accounting treatment (Ind AS 37). High-risk cases must be provisioned.
"Deemed Owner" Protection	Criminal Liability: The Board of Directors failing to nominate a specific "Agent" or "Manager" under the Mines Act, making all Directors criminally liable for accidents.	Form 29 Verification: Check if Form 29 (Nomination of Owner/Agent) has been filed with DGMS. Ensure the Board minutes reflect that safety budgets requested by the Mine Manager were *never rejected* by Finance.
Forest Rights Act (FRA)	Title Defect: Mining on forest land without the consent of the "Gram Sabha" (Village Council), which can stall the project indefinitely.	Resolution Audit: Verify the "Gram Sabha Resolution". Check if the signatures of villagers are genuine. Fake resolutions are a common cause of project cancellation.

26.5. Common Audit Observations

1. The "Job for Land" Scam: The R&R policy promises a job to one member of the displaced family. Auditors find jobs given to individuals with *no blood relation* to the land-owner, often recommended by unions.

2. CSR "General Donation": Large sums booked as CSR under "Promotion of Culture" which are actually donations to local festivals (Dussehra/Diwali) to appease local strongmen. This is not eligible CSR under the Companies Act.
3. Title Defect: Mining operations started on land where the "Mutation" (Revenue Record update) is still pending. Legally, the company does not own the land yet.

26.6. Key Frauds & Indicators

Fraud 1: The "Structure" Compensation Fraud

- The Concept: R&R pays for the land AND any "Structures" (Houses/Wells) on it.
- The Fraud: Once the land acquisition notification is issued, locals overnight build temporary mud huts or dig fake wells on the land to claim "Structure Compensation."
- Indicator: Satellite imagery (Google Earth) showing empty land in Jan and 50 houses in Feb, just before the survey.

Fraud 2: The "CSR Recycle"

- The Mechanism: The company donates ₹1 Crore to a Trust for "Skill Development."
- The Fraud: The Trust conducts a fake workshop (spending ₹10 Lakhs) and returns ₹80 Lakhs in cash to the decision-maker via shell companies.
- Indicator: CSR Partners with no website, no annual report, and "Education/Training" as the primary activity (easiest to fake).

26.7. Management Reporting (MIS) Review

The Company Secretary/Legal Head must present:

1. Land Possession Matrix: Total Required vs. Paper Possession vs. Physical Possession.
2. Litigation Hotspots: Which land parcels are under "Stay Order"?

3. CSR Burn Rate: Are we bunching 80% of CSR spend in March (poor planning)?
4. Community Grievance Index: Trends in villager complaints.

Chapter 28: IT, OT Security & Disaster Recovery

This chapter addresses the "Nervous System" of the mine. Modern mining is Industry 4.0—connected, automated, and data-driven. This creates a new risk frontier: Cyber-Physical threats. A hacker in a basement can now stop a 200-Tonne dumper or shut down a ventilation fan, risking lives.

28.1. Process Overview & Objectives

The audit of Technology in mining is split into two worlds:

1. IT (Information Technology): The corporate network (SAP, Email, HR, Finance). Objective: Data Confidentiality & Integrity.
2. OT (Operational Technology): The industrial network (SCADA, DCS, PLCs, Fleet Management). Objective: Physical Safety & Availability.

The Value Chain: Data Creation (Sensor/User) > Network Transmission > Processing (ERP/SCADA) > Storage (Database) > Backup/DR

Process KPIs (Management Dashboard):

- Uptime %: Availability of Critical Systems (FMS/SAP).
- Patch Compliance: % of servers with latest security updates.
- RTO (Recovery Time Objective): Time taken to restore operations after a crash.
- Phishing Susceptibility Rate: % of employees who clicked a fake link during a drill.

28.2. The "Why" Box: Auditor's Mindset

Auditor's Mental Model:

"I am not looking for a missing printer. I am looking for the Kill Switch.

- The Convergence Risk: Historically, IT (Office) and OT (Pit) were separate. Now they are connected. If a geologist clicks a phishing email, can the malware travel across the bridge and shut down the Crusher?

- The Ransomware Nightmare: Mining Operations cost multiple crore per hour. If ransomware locks the Fleet Management System (FMS), the mine stops. The ransom demand is calculated based on this downtime cost.
- The Goal: Verify the 'Air Gap' between IT and OT, and ensure the 'Backup' is immutable (cannot be deleted by hackers)."

28.3. Operational Data Integrity (Mine & Plant Systems)

For a modern mine, data itself is an asset class – it drives revenue, royalties, ESG and bonuses.

- Critical OT Systems Inventory
    - Audit Check: List and risk-rank OT systems that generate financial/ESG numbers: FMS, weighbridges, LIMS, CPP DCS, ETP PLCs, TSF instrumentation.
    - Control Objective: Ensure change management, backup and cyber-hardening are proportionate to data importance.
- Cross-System Reconciliations
    - Audit Check: Review daily/weekly reconciliations between OT data and ERP (tonnes, grades, fuel, power, water) and investigate thresholds for exception reporting.
    - Red Flag: Manual "adjustment journals" regularly posted to align ERP with management targets rather than OT records.

28.4. Common Industry Jargon

- SCADA (Supervisory Control and Data Acquisition): The software used to control industrial processes (Plant/Crusher).
- PLC (Programmable Logic Controller): The hardware "brain" of a machine.
- Air Gap: Physical isolation between the secure industrial network and the public internet.

- RPO (Recovery Point Objective): How much data can we afford to lose? (e.g., "Last 1 hour").
- RTO (Recovery Time Objective): How fast must we be back up? (e.g., "Within 4 hours").
- SOC (Security Operations Center): The team monitoring cyber threats 24/7.

28.5. Key Regulatory & Technical References

- Standards:
  - ISO 27001: Information Security Management.
  - IEC 62443: The Global Standard for OT/Industrial Security (Crucial for mining).
  - NIST Cybersecurity Framework: Identify, Protect, Detect, Respond, Recover.
- Statutes (Indian/Global):
  - IT Act, 2000 (Section 43A): Liability for negligence in handling sensitive data.
  - CERT-In Guidelines: Mandatory reporting of cyber incidents within 6 hours.

28.6. Detailed Audit Work Program: IT & OT Security

[Focussing on Risks from a CEO's point of view rather than a CTO's point of view]

Section 1: IT General Controls (Financial Fraud & Integrity)

Focus: Preventing cash leakage and protecting the "Single Source of Truth" (ERP).

Sub Process	CEO-Level Risk (Financial/Ops)	Audit Procedure (Field Steps)
Segregation of Duties (SOD)	Cash Theft: A single employee having the power to create a fake vendor and release a payment to them, leading to direct financial loss without detection.	The "Super User" Matrix: Analyze the ERP User Access Matrix. Identify users who have conflicting T-Codes (e.g., "Create Vendor" + "Post Invoice" + "Approve Payment"). Any user with this combination is a walking fraud risk.
Master Data Integrity	Payment Diversion: An insider changing a legitimate vendor's bank account number to their own just before a payment run, diverting millions.	Audit Trail Review: Extract the "Master Data Change Log" for the Vendor Master. Look for changes to "Bank Account Details" made just before the weekly payment run, followed by a reversal back to the original account.
Leaver Management	Data Theft & Sabotage: Disgruntled ex-employees retaining access to the network, allowing them to steal proprietary geological data (valuation loss) or delete critical financial records.	"Kill Switch" Test: Cross-match the "Resigned Employees List" (HR) from the last 90 days against the "Active Directory Users" (IT). If an ex-CFO or ex-Geologist still has an active login, it is a critical vulnerability.

Section 2: OT Security (Operational Continuity & Asset Safety)

Focus: Preventing the "Kill Switch" scenario where production stops.

Sub Process	CEO-Level Risk (Financial/Ops)	Audit Procedure (Field Steps)

The "Air Gap" (IT-OT Segregation)	Total Production Halt: A ransomware virus from a reception PC (IT) jumping to the Crusher Control System (OT), locking up the machinery.	The "Browser Test": Visit the Crusher or Mill Control Room. On the SCADA PC, try to open Google or YouTube. If the internet is accessible, the "Air Gap" is broken, and the plant is exposed to global hackers.
USB Port Hygiene	Asset Destruction: A virus introduced via a maintenance engineer's USB drive causing the Turbine or Mill to over-speed and self-destruct (Massive Capex replacement cost + months of downtime).	Physical Port Audit: Inspect the PLCs (Programmable Logic Controllers) and Operator Stations. Are USB ports physically blocked with locks or disabled via software? Unrestricted ports are the #1 entry point for sabotage (e.g., Stuxnet).
Default Passwords	Operational Hijack: A hacker taking control of the ventilation fans or pumps because they used the factory default password ("admin/1234"), putting underground lives at risk and stopping the mine.	Login Attempt: Ask the Plant Manager to log in to the DCS/PLC. Check if they are using default credentials like "admin/admin" or "1234." If yes, anyone with the equipment manual can hijack the mine.

Section 3: Disaster Recovery (Business Survival)

Focus: How fast can we bill customers again after a crash?

Sub Process	CEO-Level Risk (Financial/Ops)	Audit Procedure (Field Steps)

The "Restore" Capability	Bankruptcy Risk: Inability to restore financial data after a server crash, leading to a loss of "Receivables" records (Who owes us money?).	The "Blind Restore" Test: Do not ask *if* they backup. Ask IT to restore a specific invoice or file from 3 months ago in front of you. If they say "The tape is corrupted" or "Cloud link is down," the backup is worthless.
Ransomware Resilience	Ransom Extortion: Backups being encrypted along with the main server because they were connected to the same network. The company is forced to pay millions in ransom to get data back.	Immutability Check: Verify if backups are stored on an "Immutable" (Read-Only) media or an "Air-Gapped" tape drive that cannot be overwritten by admin credentials. If the Admin PC gets infected, it shouldn't be able to wipe the backup.
Business Continuity Drill	Paralysis: Management panicking during a real crisis because the "Disaster Recovery Plan" was only on paper and never tested.	Drill Evidence: Review the minutes of the last "Mock Disaster Drill." Did they physically switch operations to the DR Server? If it was just a "Table-Top Discussion," the actual RTO (Recovery Time) is unproven.

28.7. Common Audit Observations

1. The "Shared" Password: The Weighbridge software login is stuck on a post-it note on the monitor. All 3 shifts use the same ID ("WB_Operator"). No accountability for who manipulated the weight.
2. Pirated Software: The Survey team using cracked versions of AutoCAD or Datamine. This creates legal risk and opens backdoors for malware.

3. The "Vendor" Loophole: The mine has strict firewalls, but the "OEM Maintenance Team" (e.g., Crusher Vendor) has a permanent remote connection (TeamViewer/AnyDesk) into the system that is unmonitored.

28.8. Key Frauds & Indicators

Fraud 1: The "Logic Bomb" (Sabotage)

- The Concept: A disgruntled IT administrator plants a script to delete data *after* they leave.
- The Mechanism: "If User 'John' is removed from Active Directory, wait 30 days, then delete Table 'Vendor Master'."
- Indicator: Unexplained script files in the scheduled task scheduler or sudden data loss exactly 1 month after a layoff.

Fraud 2: Master Data Manipulation (The Bank Swap)

- The Concept: Diverting vendor payments to a personal account.
- The Mechanism:
    1. Hacker (or Insider) gains access to ERP Master Data.
    2. Changes Vendor X's Bank Account number to their own.
    3. Waits for the Pay Run.
    4. Changes the Bank Account number *back* to the original immediately after payment.
- Indicator: Audit Log (Audit Trail) showing changes to "Vendor Bank Details" just before a payment run, followed by a reversal.

28.9. Management Reporting (MIS) Review

The CIO/CISO must present:

1. Backup Success Rate: % of backups completed successfully vs. failed.
2. Vulnerability Ageing: How many "High Risk" vulnerabilities remain unpatched > 30 days?

3. Incident Report: Number of malware blocks/intrusion attempts.
4. User Access Review: Confirmation that Dept Heads have reviewed and signed off on their team's access rights this quarter.

Chapter 29: Mine Closure & Site Restoration

This chapter addresses the "Long Tail" risk of mining. The objective is to ensure that the Asset Retirement Obligation (ARO) is accurately provisioned and that the cash set aside in the Escrow Account is sufficient to cover the final cleanup, preventing a situation where the company declares bankruptcy and leaves the government with the mess.

29.1. Process Overview & Objectives

Mine Closure is not an event that happens at the end; it is a process that starts on Day 1. The objective is to return the land to a safe, stable, and productive state.

- Progressive Closure: Restoration done *during* operations (e.g., backfilling mined-out patches).
- Final Closure: Decommissioning the plant and reclaiming the final pit *after* mineral exhaustion.

The Value Chain: Life of Mine (LOM) Planning > Progressive Reclamation (Backfilling) > Financial Assurance (Escrow) > Final Decommissioning > Post-Closure Monitoring

Process KPIs (Management Dashboard):

- Reclamation Ratio: Hectares Reclaimed / Hectares Disturbed (Target > 0.8).
- Escrow Adequacy: Current Escrow Balance / Estimated Closure Cost.
- Biological Density: Number of surviving saplings per hectare on dumps.
    - ARO Variance: Actual Restoration Cost vs. Provision created in Books.

29.2. The "Why" Box: Auditor's Mindset

Auditor's Mental Model:

"This is the classic 'Kick the Can Down the Road' risk.

- The Manager's Bias: Current managers want to maximize profit *today*. Spending money on filling a hole (that won't be closed for 20 years) destroys their annual bonus.
- The Financial Risk: If the company ignores this liability, they will declare bankruptcy when the ore runs out, leaving the government (and taxpayers) to clean up the mess.
- The Auditor's Job: To ensure the Asset Retirement Obligation (ARO) on the Balance Sheet is real, and the cash in the Escrow Account is locked."

29.3. Progressive Rehabilitation & Biodiversity

Mine closure is not a one-day ceremony; it is a 10-15 year process that should start on Day-1 of mining.

- Progressive vs. Final Rehabilitation
    - Audit Check: Compare EC/Mining Plan commitments on backfilling, topsoil management and afforestation with actual annual progress (hectares rehabilitated, survival rates, maintenance spend).!
    - Risk: “Greenwashing” – tree planting events without survival tracking or protection from grazing.
- Biodiversity & Post-Mining Land-Use
    - Audit Check: Verify baseline biodiversity assessments (flora/fauna lists, critical habitats) and plans for habitat restoration, corridors and post-mining land use (water reservoirs, agriculture, solar parks).

    - Red Flag: High-biodiversity areas converted to generic monoculture plantations contrary to approved management plans.

29.4. Social Closure & Economic Transition

When the mine leaves, the town cannot disappear with it.

- Social Closure Plan
    - Audit Check: Review documented social closure plans covering workforce transition (re-deployment, training), local supplier diversification and livelihood programmes for affected communities.
    - Risk: Abrupt shutdowns leading to unrest, litigations and reputational damage that affect group-wide projects.
- Closure Funds & Governance
    - Audit Check: Where closure/rehabilitation funds are mandated (escrow, mine closure deposits), verify balances, permissible uses and actual withdrawals vs approved closure works.
    - Fraud Indicator: Using closure funds for routine opex or unrelated CSR projects under pressure just before lease expiry.

29.5. Common Industry Jargon

- PMCP: Progressive Mine Closure Plan (submitted every 5 years).
- FMCP: Final Mine Closure Plan (submitted 2 years before closure).
- Escrow Account: A specific bank account where closure funds are deposited annually, locked by the Coal Controller or IBM (Indian Bureau of Mines).

- ARO (Asset Retirement Obligation): The legal obligation associated with the retirement of a tangible long-lived asset (Ind AS 37).
- Backfilling: The process of dumping waste rock back into the mined-out pit.
- Technical vs. Biological Reclamation: Stabilizing the slope (Technical) vs. Planting trees (Biological).

29.6. Key Regulatory & Technical References

- Statutes (Indian Context):
  - MMDR Act (Section 18): Conservation and systematic development of minerals.
  - MCDR 2017 (Rule 22-27): Rules regarding PMCP and FMCP.
  - Guidelines for Preparation of Mine Closure Plan (2010): Issued by Ministry of Coal/Mines.
- Accounting Standards:
  - Ind AS 37: Provisions, Contingent Liabilities.
  - Ind AS 16 (Appendix A): Changes in Existing Decommissioning, Restoration and Similar Liabilities.
- Global Standards:
  - ICMM Principles: Integrated Mine Closure: Good Practice Guide.

29.7 Detailed Audit Work Program: Mine Closure & Site Restoration

Section 1: Financial Assurance (The Escrow Fund)

Focus: Is the statutory closure fund fully funded and locked?

Sub Process	Red Flag / Risk	Audit Procedure (Field Steps)
Escrow Calculation	Indexation Failure: Calculating the annual contribution using the	Re-Calculation: Verify the Annual Escrow Test: Check if index rate used matches the *latest* WPI-

	base rate (e.g., 2010 rates) without adding inflation (Wholesale Price Index- WPI). This results in a massive shortfall in the fund 20 years later.	indexed rate published by the Coal Controller or IBM.
Lien Verification	Unfettered Access: The company treating the Escrow as a standard "Fixed Deposit" and taking an Overdraft (OD) or Loan against it for working capital.	Bank Confirmation: Directly verify with the bank if the Escrow Account has a "Lien" marked in favor of the Regulatory Authority (Coal Controller/IBM). Ensure the company cannot withdraw funds without regulatory approval.
Area Reconciliation	Under-Reporting: Calculating liability only on the "Active Pit" area while ignoring Waste Dumps, Tailings Ponds, and Infrastructure areas to save cash flow.	EC vs. Escrow Check: Compare the "Mine Lease Area" used in the Escrow Calculation vs. the area specified in the Environmental Clearance (EC) and Consent to Operate (CTO) documents. Any gap implies under-funding.

Section 2: Physical Reclamation (Progressive Closure)

Focus: Adherence to the Progressive Mine Closure Plan (PMCP).

Sub Process	Red Flag / Risk	Audit Procedure (Field Steps)
Satellite/Drone Audit	The "Paper Forest": The Annual Report claims "5 Hectares Reclaimed," but the satellite image shows	Imagery Overlay: Overlay the PMCP Map (Plan) with current Google Earth/Drone imagery (Actual). Test: Measure the green cover. If the claimed "Reclaimed Area"

	a barren grey waste dump.	looks brown/grey, the report is false.
Topsoil Management	Resource Destruction: Dumping waste overburden on top of the fertile topsoil layer, destroying the ability to re-vegetate the land later.	Stack Inspection: Check the "Topsoil Stack" register and visit the site. Is the topsoil stored separately with proper signage and protective vegetation? If it is buried under fresh waste, raise a major non-compliance.
Dump Stability	Landslide Risk: Steep dumping (angle of repose > 37 degrees) to save space, risking landslides during the monsoon.	Slope Measurement: Use a clinometer or drone to measure the slope angle of the reclaimed dump. It must be < 28 degrees (or as per Scientific Study) to ensure long-term stability and prevent erosion.

Section 3: Accounting Provision (ARO- Ind AS 37)

Focus: True and Fair view of the Long-Term Liability.

Sub Process	Red Flag / Risk	Audit Procedure (Field Steps)
Discount Rate Test	Liability Shrinkage: Using a high discount rate (e.g., 12% Cost of Equity) to artificially shrink the Present Value (PV) of the closure liability.	Rate Validation: Review the discount rate used in the ARO model. Under Ind AS 37, it should be the Risk-Free Govt Bond Rate (e.g., 7%) for the relevant tenure. Using a higher rate suppresses the liability.
Unwinding of Discount	Profit Inflation: Failing to book the "Interest on ARO" (Unwinding cost) annually, thereby inflating profits.	P&L Scrutiny: Check "Finance Costs" in the P&L. Is there a line item for "Unwinding of Discount on Mine Closure Provision"? This represents the Time Value of

		Money increasing the liability each year.
Salvage Value	Under-Provisioning: Assuming the plant will be sold as a "Going Concern" (High Value) at the end of mine life to offset closure costs, rather than as "Scrap" (Low Value).	Assumption Check: Review the "Scrap Value" assumption in the closure model. Is it based on current scrap steel rates? Aggressive salvage value assumptions reduce the provision needed today and are a form of earnings management.

29.8. Common Audit Observations

1. The "Working Capital" Escrow: The interest earned on the Escrow Fixed Deposit is credited to the company's Current Account and used for paying salaries. *Correction:* Interest must be reinvested into the Escrow corpus.
2. The "Pot Plant" Reclamation: The company plants fast-growing invasive species (like Eucalyptus) just to show "Green Cover" quickly, destroying the local water table and biodiversity.
3. Ghost Backfilling: Contractors bill for "Backfilling" (hauling waste *back* into the pit), but GPS traces show they just dumped it at the nearest edge.

29.9. Key Frauds & Indicators

Fraud 1: The "Life of Mine" Extension (Liability Shrinkage)

- The Concept: The closure provision is discounted over the remaining life of the mine.

- The Fraud: Management arbitrarily increases the "Remaining Life" from 10 years to 20 years (by claiming "Inferred Resources" as "Reserves").
- The Impact: Increasing "Years" drastically reduces the Present Value (Liability) and the annual charge to P&L.
- Indicator: A sudden update to the Mine Plan increasing longevity without any new exploration drilling or feasibility study.

Fraud 2: The "Subsidiary Dump" (Bankruptcy Strategy)

- The Concept: The "Polluter Pays" principle.
- The Fraud: A large conglomerate moves the mining lease into a separate "Special Purpose Vehicle" (SPV) with no other assets. When the mine is exhausted, they declare the SPV bankrupt, leaving the government with the closure cost while the parent company keeps the historical profits.
- Indicator: Inter-corporate loans stripping cash out of the mining entity; Parent company refusing to give a "Corporate Guarantee" for the mine closure.

29.10. Management Reporting (MIS) Review

The Finance Controller & Environment Head must monitor:

1. Escrow Corpus Health: Is the fund growing in line with inflation?
2. PMCP Variance: Planned vs. Actual Backfilling (m^3).
3. Retrenchment Liability: Estimated cost of laying off the workforce at closure (not usually covered in the Escrow).
4. Bond Release: Has the Govt released any portion of the guarantee against completed reclamation? (Proof of quality).

Appendix

Appendix-I: List of Abbreviations & Industry Jargons

1. Regulatory Bodies & Statutory Terms

- ASP (Average Sale Price): The monthly benchmark price published by the Indian Bureau of Mines (IBM), used as the base value for calculating Ad Valorem Royalty payments.
- BEE (Bureau of Energy Efficiency): The statutory body that regulates energy consumption norms (under the PAT Scheme) for power plants and heavy industries.
- CCO (Coal Controller Organization): The regulatory body governing coal mines in India, performing a role similar to the IBM for non-coal minerals.
- CTO (Consent to Operate): A periodic license issued by the State Pollution Control Board allowing a mine or plant to operate within specific pollution limits.
- DGMS (Directorate General of Mines Safety): The regulatory agency under the Ministry of Labour responsible for enforcing safety laws and investigating accidents in mines.
- DMF (District Mineral Foundation): A statutory trust funded by miners (paying an additional 10% or 30% of Royalty) to be used for the benefit of mining-affected communities.
- EC (Environmental Clearance): The master permission issued by the MoEF&CC specifying the maximum production capacity and environmental conditions for the project.
- IBM (Indian Bureau of Mines): The technical regulator for non-coal minerals in India; responsible for approving Mining Plans and monitoring mineral conservation.
- MMDR Act (Mines and Minerals Development and Regulation Act): The principal legislation governing the mining sector in India.

- NMET (National Mineral Exploration Trust): A statutory levy (2% of Royalty) collected from miners to fund regional exploration activities.
- PESO (Petroleum and Explosives Safety Organization): The regulator overseeing the storage, transport, and handling of explosives and petroleum products.

2. Mining Operations & Geology (The Pit)

- BCM (Bank Cubic Meter): The volume of rock or soil in its natural, in-situ state (solid ground) before it is excavated. It is the standard unit for billing excavation work.
- G-Axis (G1, G2, G3, G4): Codes under the UNFC system representing the level of geological confidence. G1 represents "Proven/Detailed" reserves (High Confidence), while G4 represents "Reconnaissance" (Low Confidence).
- HEMM (Heavy Earth Moving Machinery): Large equipment used in mining operations, such as Excavators, Dumpers, Dozers, and Graders.
- LCM (Loose Cubic Meter): The volume of material after it has been blasted or dug up; it expands due to air voids (Swell Factor) compared to BCM.
- LOM (Life of Mine): The total estimated time period in which the ore reserves will be extracted until the mine is exhausted.
- OB (Overburden): The waste rock or soil lying above the ore body that must be removed to access the valuable mineral.
- PF (Powder Factor): A key efficiency metric indicating the amount of explosives (in kg) used to break one tonne or cubic meter of rock.
- ROM (Run of Mine): Raw, unprocessed ore directly from the blasting face before any crushing, screening, or treatment.
- SR (Stripping Ratio): The ratio of Waste removed to Ore extracted (e.g., 1:3 means 3 tons of waste for 1 ton of ore). It is a major cost driver.

- TKPH (Tonne-Kilometre Per Hour): A rating for OTR tires that measures the maximum workload (Load X Speed) the tire can handle without overheating.

3. Processing, Plant & Quality

- DMT (Dry Metric Tonne): The weight of ore excluding moisture content. Sales realization and pricing are often based on DMT.
- GCV (Gross Calorific Value): The energy content of coal (kCal/kg). Coal pricing is primarily determined by GCV bands.
- MIC (Metal In Circuit): The estimated value of partially processed ore or slurry held inside the plant's tanks and pipes (Work in Progress).
- Recovery Rate: The percentage of valuable metal successfully extracted from the ore during processing (e.g., 90% recovery means 10% was lost in tailings).
- Tailings: The waste slurry (water + crushed rock) discharged from the beneficiation plant after the valuable mineral has been extracted.
- WMT (Wet Metric Tonne): The weight of ore including its natural moisture content. Transport is usually billed on WMT.
- ZLD (Zero Liquid Discharge): An environmental mandate ensuring no industrial wastewater is discharged outside the plant premises; all water must be recycled.

4. Power & Energy

- CPP (Captive Power Plant): A power plant set up by an industry primarily for its own consumption rather than for selling power to the grid.
- ESCerts (Energy Saving Certificates): Tradable certificates earned by efficient plants under the PAT scheme, or purchased by inefficient ones to meet targets.

- PAT (Perform, Achieve and Trade): A regulatory scheme by the BEE to reduce specific energy consumption in energy-intensive industries.
- SHR (Station Heat Rate): The efficiency of a power plant, measured in kCal of fuel required to generate 1 kWh of electricity.
- SOC (Specific Oil Consumption): The amount of secondary fuel (oil) used to support combustion in a boiler, measured in ml/kWh.

5. Logistics & Commercial

- CC (Carrying Capacity): The maximum permissible load a railway wagon can carry. Freight is often charged on CC even if the actual load is lower.
- Demurrage: Penalty charged by Railways or Transporters for detaining the vehicle or rake beyond the allowed "Free Time" for loading/unloading.
- L1 (Lowest Bidder): The vendor who quotes the lowest price in a tender process. This is often the default selection criterion in PSU/Indian procurement.
- MDO (Mine Developer and Operator): A contracting model where a private party takes full responsibility for land acquisition, mining, and production in exchange for a mining fee per tonne.
- PVC (Price Variation Clause): A contractual formula that adjusts rates based on market fluctuations in diesel, labor, or material indices.
- RCM (Reverse Charge Mechanism): A GST provision where the recipient of goods/services (e.g., Mine Owner paying Royalty) is liable to pay the tax instead of the supplier.
- TCO (Total Cost of Ownership): A procurement evaluation method that considers not just the buying price, but also operating (fuel/maintenance) costs over the asset's life.

6. Finance & Accounting

- ARO (Asset Retirement Obligation): A legal and accounting obligation (Ind AS 37) to restore the mine site after closure, for which a financial provision is created.
- CWIP (Capital Work in Progress): The value of assets currently under construction or installation that have not yet been capitalized into the Fixed Asset Register.
- IFRIC 20: An international accounting interpretation that guides when to capitalize "Stripping Costs" (waste removal) during the production phase.
- NRV (Net Realizable Value): The estimated selling price of inventory minus the cost to complete and sell. Inventory is valued at the Lower of Cost or NRV.
- PCFC (Packing Credit in Foreign Currency): A low-interest export finance loan used to fund working capital requirements prior to shipment.

7. IT, OT & ESG

- Air Gap: A security measure where the industrial network (OT) is physically isolated from the corporate network (IT) and internet to prevent hacking.
- DCS (Distributed Control System): A computerized control system for a process or plant with many control loops (e.g., Power Plant).
- FMS (Fleet Management System): Software and hardware (GPS) used to track mining trucks, optimize routes, and prevent theft.
- LTIFR (Lost Time Injury Frequency Rate): A standard safety metric measuring the number of lost-time injuries occurring per million man-hours worked.
- OT (Operational Technology): The hardware and software that detects or causes a change through the direct monitoring/control of physical devices (e.g., PLCs, SCADA).
- SCADA (Supervisory Control and Data Acquisition): System used to monitor and control industrial processes and machinery remotely.

Appendix 2- Chapter-wise Initial Data Requirement List

This list is designed to be sent to the Auditee (Client) prior to the start of the audit fieldwork. It covers specific operational documents, statutory registers, and system reports required to execute the Audit Work Programs detailed in the book.

Part I: The Foundation (Strategy & Regulations)

Chapter 1: The Indian Mining Landscape

- Strategic Business Plan: Approved 3-5 year plan covering production targets, stripping ratios, and cost projections.
- Cost of Production (COP) Report: Monthly MIS showing the breakdown of Variable (Fuel, Explosives) vs. Fixed Costs.
- Competitor Benchmarking: Internal reports comparing EBITDA/Tonne with peers (e.g., Vedanta, Tata, NMDC).
- Org Structure: Detailed Organogram of Mine, Plant, and Geology departments.

Chapter 2: The Regulatory Web

- Mining Lease (ML) Deed: Copy of the registered deed and any supplementary lease deeds.
- Environmental Clearance (EC): EC Letter from MoEF&CC and compliance status reports submitted half-yearly.
- Consent to Operate (CTO): Valid Air & Water Consents from the State Pollution Control Board (SPCB).
- IBM Returns: Monthly (Form F) and Annual (Form G) returns filed with the Indian Bureau of Mines.
- Royalty Challans: Proof of Royalty, DMF, and NMET payments for the audit period.

Chapter 3: Geology & Exploration

- Geological Report (GR): The base document defining Proved (G1) and Probable (G2) reserves. The year-end summary of Reserves & Resources

- Drill Core Logs: Database of exploration drilling (Lithology, Grade, Depth).
- Assay Reports: Third-party lab analysis reports for boreholes.
- Block Model Data: Access to Surpac/Datamine files showing the 3D ore body model.
- Resource Reconciliation: Report comparing "Modelled Reserve" vs. "Actual Mined" for the year.
- Drilling Contracts: Rate cards for different strata (Soft soil vs. Hard rock).
- Daily Drilling Reports (DDR): Signed by the Rig Geologist.
- Core Box Inventory: Photos/Logs of the physical core boxes stored in the "Core Shed".
- Borehole Plan: Map showing planned vs. actual drill hole locations (Coordinates).

Chapter 4: Mine Planning & Projects

- Approved Mining Plan: The 5-Year Mining Plan/Scheme approved by IBM/CCO.
- Feasibility Report (DPR): Detailed Project Report for any new expansion or shaft sinking.
- Capex Budget: Board-approved capital budget vs. Actual commitment.
- Project S-Curves: Physical and Financial progress charts for ongoing civil/infra works.
- Land Acquisition Status: Map showing Acquired Land vs. Forest Land vs. Encroachments.

Part II: Core Mining Operations (The Pit)

Chapter 5: Pre-Stripping & OB Removal

- Monthly Mine Plan: Short-term production schedules issued to the operations team.
- Survey Reports: Monthly Joint Survey Reports (JSR) signed by Surveyor and Contractor.

- Stripping Ratio (SR) Calc: Monthly calculation of OB removed (m^3) / Ore extracted (MT).
- Dump Management Plan: Map showing active waste dumps and their approved capacity.
- Contractor Agreements: Work Orders for OB removal (MDO/Contractor).

Chapter 6: Production & Extraction

- Form D (Production Register): Statutory daily log of raising, dispatch, and closing stock.
- Blasting Reports: Daily "Blast Note" showing Hole Depth, Spacing, and Explosives charged.
- Drill Logbooks: Shift-wise drilling meters recorded by operators.
- Rainfall Data: Meteorological log (Pluviometer) to verify "Force Majeure" downtime.
- Shift Boss Reports: Handwritten shift logs mentioning breakdowns or stoppages.

Chapter 7: HEMM & Maintenance

- HMR / KMR Logs: Hour Meter/Odometer readings for all heavy machinery (Dumpers, Excavators).
- Equipment Availability Report: Monthly MIS showing MTBF (Mean Time Between Failures) and MTTR.
- Diesel Issue Register: Log of High-Speed Diesel (HSD) dispensed to specific Equipment IDs.
- Tyre History Cards: Lifecycle tracking for each tyre (Serial No, Fitment Date, TKPH, Scrap Date).
- Job Cards: Maintenance records for major breakdowns (Engine/Transmission overhauls).

Chapter 8: Inventory Management

- Volumetric Survey Report: Third-party Drone/Total Station survey reports of stockpiles.
- Bulk Density Test Report: Lab certificate establishing the Conversion Factor (Volume to Weight).

- Stock Ledger: ERP dump showing Opening, Production, Dispatch, and Closing Stock.
- Non-Moving Stock List: Aging analysis of ROM and processed ore.
- Write-off Approvals: Note sheets for any stock adjustments/losses booked.

Part III: Value Addition (The Plant)

Chapter 9: Ore Beneficiation

- Mass Balance Sheet: Input (Feed) = Output (Concentrate + Tailings) reconciliation.
- Assay Lab Reports: Daily grade analysis of Feed, Concentrate, and Tailings.
- Yield Report: Plant Recovery % vs. Design Yield.
- Reagent Consumption Log: Daily usage of Frothers, Collectors, and Flocculants.
- Metal Accounting Report: Reconciliation of "Metal in Circuit" (MIC).

Chapter 10: Captive Power Plant (CPP)

- Generation Report: Daily Gross Generation vs. Net Export (after Aux consumption).
- Coal Analysis Report: "As Fired" GCV and Moisture reports.
- Specific Metrics: Monthly calculation of Station Heat Rate (SHR) and Specific Oil Consumption (SOC).
- Grid Bills: Electricity Export/Import bills from the State Grid.
- Ash Utilization: Fly ash disposal/sales register.

Chapter 11: ETP & Water Management

- Water Balance Diagram: Schematic showing fresh water intake vs. recycled water.
- Flow Meter Log: Daily readings of Inlet and Outlet flow meters.
- ETP Logbook: Record of chemicals used and downtime in the Effluent Treatment Plant.
- Hazardous Waste Manifest: Form 10/Manifests for disposal of sludge/used oil.

Part IV: Quality & Assurance

Chapter 12: Weighbridge Operations

- Calibration Certificates: Legal Metrology stamping for all weighbridges.
- Zero Test Log: Daily record of "Zero Error" checks performed at shift start.
- Tare Weight Register: Database of historical tare weights for registered trucks.
- Manual Override Report: System log of all weighment tickets where the weight was manually edited.
- Ticket Sequence Log: Report to identify missing or skipped ticket numbers.

Chapter 13: Quality Control (Assay Lab)

- Sample Register: Log of samples collected (Tag No vs. Location/Truck No).
- Umpire Lab Results: Comparative analysis (Mine Lab vs. Third Party Lab vs. Customer Lab).
- Calibration Log: Records of XRF/Analyzer calibration using Standard Reference Material (SRM).
- Auto-Sampler Log: Maintenance and uptime records of mechanical auto-samplers.

Part V: Supply Chain & Logistics

Chapter 14: Material Procurement

- Purchase Order (PO) Register: Dump of all POs issued during the period.
- Comparative Statements (CS): Bid evaluation sheets for top 25 high-value orders.
- Single Tender Justification: Approval notes for nomination/single-bidder orders.

- RE-7 (Explosives Return): Statutory return showing opening stock, receipt, usage, and closing stock of explosives.
- Fuel Dip Charts: Daily dip readings of HSD underground tanks.

Chapter 15: Service Contracts

- Contract Master: List of all active Service Orders (Mining, Transport, Security).
- Measurement Books (MB): The physical logbooks recording work done (e.g., m^3 excavated) before billing.
- PVC Calculation Sheet: Working files for Price Variation (Escalation) claims paid to contractors.
- Performance Bank Guarantees (PBG): Copy of active BGs held against contracts.

Chapter 16: Warehouse Management

- Slow/Non-Moving Inventory: Aging report of spares (>1 year, >3 years).
- Gate Entry Register: Security gate logs of material inward vs. GRN creation date.
- Physical Verification (PV) Report: Annual PV sheets and reconciliation of shortages.
- Scrap Sales Register: List of scrap auctions and gate passes for scrap removal.

Chapter 17 & 18: Logistics (Inward & Outward)

- Trip Sheets: Logs of trips made by internal transporters (time out vs. time in).
- Route Map: Approved distance chart used for freight payments.
- Transit Loss Report: Analysis of Loading Weight vs. Unloading Weight shortages.
- Railway Receipts (RR): Documents showing chargeable weight vs. actual weight for rail logistics.
- E-Way Bills: GST E-Way bill dump for all dispatches.

Part VI: Human Capital & Admin

Chapter 19: Payroll & Incentives

- Biometric Dump: Raw attendance data vs. Shift Roster.
- Form IV (Overtime Register): Statutory log of OT hours worked.
- Incentive Policy: Signed policy document for Production Bonus/Incentives.
- Contract Labor Compliance: PF/ESI challans and Form V (Abstract of Contractor Labor).

Chapter 20: Administration

- Vehicle Logbooks: Usage logs (Odometer Start/End) for hired vehicles.
- Guest House Register: Occupancy records vs. Grocery/Food bills claimed.
- AMC Contracts: Annual Maintenance Contracts for IT, ACs, and Office facilities.

Part VII: Finance, ESG & IT

Chapter 22: Accounting & Closure

- Trial Balance: Detailed GL dump with sub-ledgers.
- Provision Schedule: List of all expenses accrued but not paid (OB removal, Power).
- Actuarial Valuation: Report for Gratuity and Leave Encashment liabilities.
- Related Party Transactions (RPT): Register of contracts with group companies (Form MBP-4).
- Monthly Cost Sheets: Showing Cost per Tonne breakdown.
- Tax Returns: GSTR-3B, GSTR-1, and Income Tax Computation.
- Legal Case Status: List of disputes with Mining Department/Pollution Control Board.

Chapter 23: Treasury & Hedging

- Bank Reconciliation Statement (BRS): For all active bank accounts.
- Sanction Letters: Loan agreements listing covenants and interest rates.

- Hedging Register: List of open Forex forward contracts and MTM valuation.
- Authorized Signatory List: Board resolution approving bank signatories.
- Treasury Policy: The Board-approved document defining risk limits (e.g., "No more than 50% of exposure can be unhedged").
- Forex Blotter: List of all open forward/option contracts.
- Credit Rating Report: Latest report from CRISIL/Fitch/Moody's.

Chapter 24: Fixed Assets

- Fixed Asset Register (FAR): Detailed list of assets with capitalization dates and depreciation rates.
- CWIP Ageing: List of open capital projects and their start dates.
- Capitalization Note: Engineering certificate for assets capitalized during the year.
- IFRIC 20 Calculation: Stripping cost capitalization working model.
- Capitalization Policy: Threshold limit (e.g., items <1000 are expensed).
- Geological Report: Showing the "Stripping Ratio" for the Life of Mine (LOM).
- Physical Verification (PV) Report: Internal team's findings.
- Used Oil/Scrap Sale Register: To trace disposal of old engines/parts.

Part VIII: Sustainability, Governance, Risk & Compliance

Chapter 25: ESG & Sustainability

- BRSR Report: Draft Business Responsibility and Sustainability Report.
- GHG Inventory Calculation: Workings for Scope 1 (Diesel/Coal) and Scope 2 (Grid Power) emissions.
- Safety Statistics: LTIFR (Lost Time Injury Frequency Rate) calculation and accident logs.
- CSR Projects: List of projects, implementing agencies, and utilization certificates.

- Energy Bills: Electricity and Diesel purchase registers (Chapter 14 data).
- Accident Register: Form 4 (DGMS) for safety incidents.
- Water Cess Returns: Filings made to the Pollution Control Board.

Chapter 26: Statutory Compliance

- Compliance Dashboard: Status of all statutory filings (Tax, Labour, Environment).
- Show Cause Notices (SCN): File of notices received from DGMS, IBM, or PCB.
- Safety Committee Minutes: Minutes of the monthly Pit Safety Committee meetings.
- Form 4A: Notice of accidents sent to DGMS.
- Statutory Appointments List: List of Managers, Surveyors, Overmen appointed under the Act.
- Accident Register: Form J / Form K.
- Pollution Monitoring Reports: Last 12 months (Air/Water/Noise).
- DGMS Violations: Copies of the last 3 inspection letters from DGMS officials.

Chapter 27: Secretarial, CSR & Legal (Land R&R)

- Land Schedule: List of all plots (Survey Numbers), Area, and Status (Freehold/Leasehold).
- R&R Policy: The Board-approved compensation benefits package.
- CSR Project List: Approved projects for the year.
- Legal Case Tracker: List of all pending suits (Civil/Criminal).
- Minutes of Meetings (MOM): Board and CSR Committee minutes.

Chapter 28: IT & OT Security

- User Access Matrix: List of ERP users and their T-Codes (to check Segregation of Duties).
- Backup Logs: History of successful/failed backups for ERP and SCADA.
- VAPT Report: Vulnerability Assessment & Penetration Testing report.

- Disaster Recovery Plan: Copy of the DRP and results of the last mock drill.
- Network Topology Diagram: Showing the connection (firewalls) between IT and OT zones.
- User Access Dump: Eg: List of users with "SAP_ALL" or "Super Admin" rights.
- DR Drill Report: Minutes of the last mock disaster drill.
- Backup Policy: Schedule of backups (Daily/Weekly) and location (On-prem/Cloud)

Chapter 29: Mine Closure

- Escrow Passbook: Bank statement of the Mine Closure Escrow Account.
- Progressive Mine Closure Plan (PMCP): Copy of the approved reclamation plan.
- ARO Calculation Model: Excel file showing the provision calculation (Discount rate, Inflation, Life of Mine).
- Survey Report: Showing "Area Disturbed" vs. "Area Reclaimed" for the year.
- Actuarial Report: Discount rates and inflation assumptions used for the liability.

Appendix-3: Moving "Random" to "Risk-Based" Samples

Use data patterns to pinpoint anomalies, ensuring the audit time is spent verifying only the highest-risk transactions. Below procedures can be used to pick high risk samples for the audit procedures mentioned in the chapters.

Part I: The Foundation (Strategy, Geology & Planning)

Chapter 1 & 2: Strategy & Regulatory Compliance

Objective: Detect illegal production capacity breaches and Royalty underpayments.

Data Source	Analytical Procedure (The Logic)	High-Risk Sample Selection Criteria
Daily Production Log	EC Limit Stress Test: The Environmental Clearance (EC) has a strict annual limit. *Logic:* Divide Annual EC Limit by 330 working days to get "Max Daily Capacity."	Select dates where Daily Production > 120% of Rated Capacity. *(Consistently exceeding daily limits is a violation of the Air Act/Consent to Operate).*
Royalty Returns (Form G)	The "IBM Price" Check: Royalty is often paid on the *Higher* of Invoice Value or IBM Average Sale Price (ASP). *Logic:* Compare Sales Realization/Tonne vs. IBM ASP/Tonne for that month.	Select months where Invoice Price < IBM ASP. *(Risk: The company paid Royalty on Invoice Price instead of the higher IBM Price, leading to statutory non-compliance).*

Competitor Data	Margin Erosion Test: Compare Company EBITDA/Tonne vs. Peer Group (e.g., Vedanta/NMDC/Tata).	Select quarters where Company EBITDA is < 80% of Industry Average. *(Indicates massive cost leakage or under-invoicing of sales).*

Chapter 3: Geology & Exploration (The Resource Risk)

Objective: Detect "Resource Inflation" (fake reserves) and Model failures.

Data Source	Analytical Procedure (The Logic)	High-Risk Sample Selection Criteria
Block Model vs. Production	Model Reconciliation Factor (F1): *Formula:* Actual Grade Mined / Block Model Predicted Grade. Ideal ratio is 1.0.	Select Mining Blocks/Benches where Variance is > ±15%. *(If Actual < Model, the Reserve Valuation is inflated. If Actual > Model, the Geologist is incompetent or under-reporting).*
Assay Database	The "Spike" Filter: Search for anomalous high-grade samples in the database.	Select drill holes showing Grade > 3x Average Deposit Grade. *(Check if these "nuggets" were capped. If not, they artificially inflate the entire deposit value).*

Chapter 4: Mine Planning & Projects

Objective: Detect "High Grading" (Sustainability risk) and Project padding.

Data Source	Analytical Procedure (The Logic)	High-Risk Sample Selection Criteria
Mine Plan vs. Actuals	Stripping Ratio (SR) Deviation: *Logic:* Compare Planned SR vs. Actual SR. *(Removing waste is a cost; mining ore is revenue).*	Select months where Actual SR << Planned SR (e.g., Plan 1:3, Actual 1:1.5). *(This is "High Grading"—picking easy profits today and leaving massive waste removal liability for the future/bankrupting the mine).*

Project S-Curve	Physical vs. Financial Progress: Compare % Money Spent vs. % Work Done.	Select Projects where Financial Progress > Physical Progress + 10%. *(Indicates advances paid to contractors without work being delivered—a common kickback mechanism).*
DPR (Feasibility Study)	IRR Sensitivity Test: Review the project's Internal Rate of Return (IRR) model.	Select projects where IRR turns Negative with a 10% drop in Commodity Price. *(These are high-risk "Vanity Projects" that should probably be shelved).*

Part II: Core Mining Operations (The Pit)

Chapter 5 & 6: Production & Excavation

Objective: Detect Over-Reporting of Production or OB Removal.

Data Source	Analytical Procedure (The Logic)	High-Risk Sample Selection Criteria
Fleet Management System (FMS)	The "Cycle Time" Logic: Calculate the average cycle time (Load to Dump to Return) for each dumper.	Filter trips where Cycle Time < 50% of Standard. (e.g., A 20-min trip completed in 5 mins implies "Ghost Trips" logged manually).
Survey Reports	The "Swell Factor" Logic: Calculate Ratio of Loose Cubic Meter (Truck Volume) to Bank Cubic Meter (Survey Volume).	Select months where LCM/BCM Ratio > 1.3 (Standard is ~1.1 to 1.2). High ratio implies loose packing or inflated truck counts.
Explosives Log	Powder Factor (PF) Analysis: Calculate Kg	Select Blast IDs where PF > 20% of Benchmark. (High explosives

	Explosives / Tonne Broken for each blast.	use suggests theft or inefficient blasting).

Chapter 7: HEMM & Maintenance

Objective: Detect Fuel Theft and Fake Spare Parts consumption.

Data Source	Analytical Procedure (The Logic)	High-Risk Sample Selection Criteria
Fuel Issue Register	Consumption Deviation: Calculate Liters/Hour for each machine ID. Plot a standard deviation curve.	Select machines where consumption is > 2 Standard Deviations above the fleet average. (Potential fuel theft from specific vehicles).
Spare Parts Store	The "Lifecycle" Test: Calculate the interval (hours) between two replacements of the same part (e.g., Battery, Tyre).	Select Part IDs where Replacement Interval < 50% of Manufacturer's Warranty life. (e.g., Battery replaced in 3 months vs 12 months).
HMR Log	Utilization Anomaly: Compare Engine On Hours vs. Working Hours.	Select days where Idle Hours > 40% yet fuel consumption remains normal.

Part III: Value Addition (The Plant)

Chapter 9 & 10: Beneficiation & Power

Objective: Detect Yield Manipulation and Efficiency Loss.

Data Source	Analytical Procedure (The Logic)	High-Risk Sample Selection Criteria

Assay Lab Data	Mass Balance Check: Formula: Input (Metal) = Output (Concentrate) + Waste (Tailings).	Select days where Unaccounted Loss > 2%. (Metal disappearing from the circuit).
Weighbridge vs. Belt Scale	The "Calibration Drift": Compare tonnage recorded at Plant Infeed (Belt Scale) vs. Mine Weighbridge.	Select days where Variance > 1.5%. (Indicates one of the weighing systems is manipulated/faulty).
CPP Logbook	Heat Rate Analysis: Calculate Coal Consumed (kCal) per Unit of Power Generated (kWh).	Select shifts where Station Heat Rate (SHR) spikes but the Load Factor (MW) remains constant. (Indicates poor coal quality or theft).

Part IV: Quality & Assurance

Chapter 12: Weighbridge Operations (Critical)

Objective: Detect Weight Manipulation (Under-weighing sales/Over-weighing purchases).

Data Source	Analytical Procedure (The Logic)	High-Risk Sample Selection Criteria
Weighment Dump	Tare Weight Analysis: Compare the "Tare Weight" of a specific truck ID today vs. its historical average (last 30 days).	Select tickets where Tare Weight Deviation > ±300 kg. (Indicates water in tires, dual tanks, or driver sitting inside during tare).
Time Stamp Data	The "Midnight" Pattern: Analyze the frequency of	Select tickets generated between 12:00 AM to 4:00 AM (High fraud window) or during shift changeover times.

	weighments by hour of the day.	
Speed Analysis	Weighing Duration: Calculate time difference between "Gate In" and "Gate Out".	Select trucks where Turnaround Time < 5 mins (Physically impossible to enter, weigh, dump, tare, and exit).

Chapter 13: Quality Control

Objective: Detect Grade Suppression (Reporting lower grade to pay less Royalty).

Data Source	Analytical Procedure (The Logic)	High-Risk Sample Selection Criteria
LIMS Data	Grade Correlation: Plot Mine Face Grade (Geology) vs. Crusher Feed Grade (Plant) vs. Dispatch Grade (Sales).	Select periods where Dispatch Grade < Mine Grade by > 5%. (Ore grade doesn't just evaporate; high risk of under-reporting).
Sampling Log	The "Sunday" Effect: Compare average grade reported on Weekends/Holidays vs. Weekdays.	Select samples processed on Sundays/Public Holidays if the grade is consistently lower/higher (Lab staff manipulation).

Part V: Supply Chain & Commercial

Chapter 14 & 15: Procurement

Objective: Detect Split POs and Bid Rigging.

Data Source	Analytical Procedure (The Logic)	High-Risk Sample Selection Criteria
PO Dump	Split PO Check: Identify multiple POs raised to	Select POs where the Sum of split POs > Delegation of Authority

	the *same vendor* on the *same day* (or within 3 days).	(DOA) limit. (Circumventing approval limits).
Vendor Master	Benford's Law: Apply Benford's Law to the *value* of Purchase Orders.	Select PO values that appear excessively frequently (e.g., many POs for exactly 49,900 when the limit is 50,000).
Bid Data	Pattern Recognition: Compare winning bid vs. L2 and L3 bids.	Select tenders where the difference between L1 and L2 is < 1% (Suspiciously close) or > 30% (Predatory pricing).

Chapter 17 & 18: Logistics

Objective: Detect Material Theft in Transit.

Data Source	Analytical Procedure (The Logic)	High-Risk Sample Selection Criteria
GPS/FMS Data	Transit Time Analysis: Calculate Average Time taken between Mine and Port/Plant.	Select trips where Duration > Standard Time + 20%. (Truck likely stopped for theft/adulteration).
Route Replay	Geofence Breach: Analyze GPS coordinates against the approved route corridor.	Select trips that show "Route Deviation" alerts or stops at unauthorized "Dhabas" (roadside eateries).

Part VII: HR

Chapter 19: Payroll

Objective: Detect Ghost Workers and OT Fraud.

Data Source	Analytical Procedure (The Logic)	High-Risk Sample Selection Criteria

Attendance Data	Gate In vs. Shift: Compare "Gate In" timestamp with "Shift Start" time.	Select employees marked "Present" but with No Biometric Punch (Manual override).
OT Register	Fatigue Analysis: Calculate consecutive OT hours.	Select employees claiming > 4 hours OT every day for a month (Physically impossible, implies collusion).

Chapter 22 & 24: Finance & Assets

Objective: Detect Capitalization of Opex.

Data Source	Analytical Procedure (The Logic)	High-Risk Sample Selection Criteria
GL Dump (Capex)	Keyword Search: Search the narration field of Capital Additions.	Select entries containing words: "Repair", "Maintenance", "Service", "Replacement", "Paint", "Consumable". (These are likely Opex hidden as Capex).
CWIP Ledger	Aging Analysis: Calculate age of projects in Capital Work in Progress.	Select projects sitting in CWIP for > 2 Years with no movement (Hidden impairment/Abandoned projects).

A Note on Scope: To ensure this handbook remains a sharp, field-ready tool rather than a thousand-page academic volume, I have intentionally limited the coverage of generic audit checkpoints. Standard procedures common to all industries are assumed to be part of the auditor's existing toolkit, hence covered in limited manner.

Copyrighted Material

ISBN: 9789356354920

The Knowledge Engine: Comprehensive Online Risk Library

In addition to the methodologies outlined in this guide, Risk Vigilant Bharat has developed an exhaustive, cloud-based Risk Library designed specifically for modern audit professionals. This dynamic resource features comprehensive audit work programs, detailing over 15 critical analytical aspects for every single checkpoint.

As of March 31, 2026, the library houses a continuously expanding repository of over 5,000 checkpoints spanning multiple industries. Because the regulatory and operational landscape is always shifting, this database is continuously updated to reflect emerging risks and industry best practices.

Access to this living database is available through a highly accessible, nominal subscription model. To equip your team with this tool or to request further information, please reach out to **contact@riskvigilantbharat.com**.

Made in the USA
Monee, IL
05 November 2026

67809034R00157